DRY TIMES

BLUEPRINT FOR A RED LAND

MARK STAFFORD SMITH AND JULIAN CRIBB

CSIRO
PUBLISHING

National Library of Australia Cataloguing-in-Publication entry

Stafford Smith, Mark.

Dry times: blueprint for a red land / Mark Stafford Smith, Julian Cribb.

9780643095274 (pbk.)

Includes index.
Bibliography.

Arid regions ecology – Australia.
Arid regions agriculture – Australia.
Biodiversity – Australia.
Sustainable development – Australia.
Natural resources – Australia – Management.

Cribb, Julian

333.7360994

Published by
CSIRO PUBLISHING
150 Oxford Street (PO Box 1139)
Collingwood VIC 3066
Australia

Telephone: +61 3 9662 7666
Local call: 1300 788 000 (Australia only)
Fax: +61 3 9662 7555
Email: publishing.sales@csiro.au
Web site: www.publish.csiro.au

Front cover photos by Greg Rinder/CSIRO (top), iStockphoto (bottom)
Back cover photo by Mark Stafford Smith

Set in 10.5/14 Adobe Palatino and Optima
Edited by Peter Storer Editorial Services
Cover and text design by James Kelly
Typeset by Desktop Concepts Pty Ltd, Melbourne
Index by Russell Brooks
Printed in China by 1010 Printing International Ltd

CSIRO PUBLISHING publishes and distributes scientific, technical and health science books, magazines and journals from Australia to a worldwide audience and conducts these activities autonomously from the research activities of the Commonwealth Scientific and Industrial Research Organisation (CSIRO).
The views expressed in this publication are those of the author(s) and do not necessarily represent those of, and should not be attributed to, the publisher or CSIRO.

CONTENTS

ACKNOWLEDGEMENTS

The thinking in this book arises from nearly three decades of life and experiences in desert Australia, and a side interest (that has now become central) in climate change. Many years of working with the international body of scientists who study the global climate has convinced me that we face a future of rapid change. This change is almost certainly going to come upon us faster than the carefully couched public estimates, creating huge uncertainties about what the future will bring for our children and our children's children.

Deserts are all about dealing with uncertainties. They hold vital lessons of immediate relevance to a world beset by climatic, economic and resource variability – a world where the stable certainties on which our civilisation was built seem increasingly fragile.

In my time in the desert, I have valued countless generous exchanges with many desert dwellers, from pastoralists back of Bourke to traditional elders among remote desert sands, with research colleagues in dry river beds and public servants in tall city towers. These encounters have often been challenging, but invariably welcoming, and it is to all these people who care about the desert that this book is dedicated.

Reflecting on these interactions, it has often struck me that desert dwellers – myself included – are endlessly bemused by the slings and arrows that the 'world out there' seems to throw at them. Discussions start with tirades about the unfairness of drought, continue with amazement at how people in distant capitals such as Canberra have no idea what life is like in the bush, and round off with disgust that the national news only ever mentions the outback when something disastrous (outback murders) or humorously quirky (antics around the first traffic lights in Alice Springs) rears its head. 'Why can't they take us seriously?', comes the cry.

Well, I have come to see that, like drought, most of these frustrations are a more or less inevitable outcome of how the whole place works, not some evil conspiracy. Understanding this is the first step to correcting the situation as far as possible, or living with it more happily in so far as it is not. Hopefully, desert readers will emerge from the end of this book with greater equanimity and resolve than when they started. And non-desert readers will have a better appreciation of how one of our nation's principal wealth-generating regions really works.

At the same time as I was experiencing these desert encounters, I met a set of people outside the desert who are grappling with climate changes and how to adapt to live with them. Here I found another group that is bewildered by what the apparently uncontrollable forces of nature are imposing on them. 'How can we make decisions in the face of such uncertainty?', they ask, desperate for more certainty. Unfortunately, science is telling us that certainty is often not possible. Of course, desert dwellers have lived with uncertainty for eons, and so their tale is especially important.

Desert differences challenge our comfortable preconceptions, and push the limits of our assumptions. Because of this, the understanding of why the desert operates in special ways to cope with uncertainty and resource limitations is a key to our understanding of the nature of Australia as a whole, and indeed of the planet, as we engage with the dry tsunami of climate change.

The book builds on a body of work from the Desert Knowledge Cooperative Research Centre (DKCRC), but the writing was specifically enabled by substantial funding from Land & Water Australia and from CSIRO. My partnership with Julian Cribb has been supported by funding from the Commonwealth Government (then Departments of Environment and Heritage, and Agriculture Fisheries and Forestry, through the National Heritage Trust). We thank all these sponsors for their support.

In particular, the opportunity for me to sit back and contemplate a more popular synthesis of the issues of desert living arose from the generous granting of a Land & Water Australia Senior Research Fellowship to me by the Land & Water Australia Board, in conjunction with CSIRO's support. Land & Water Australia, now sadly defunct, set up these fellowships to allow selected researchers 'time out' from administration and grant chasing to produce a major work outlining their reflections on, and recommendations from, the best research in their field; I have valued this 'time out' hugely.

Behind this synthesis lie many ideas which have emerged through the DKCRC's Science of Desert Living project, and I have depended on the willingness of many DKCRC researchers to share recent ideas – too many to name individually, but they are referenced in the footnotes. I am very grateful to them all. Funding for this background work came from the Australian Government Cooperative Research Centres Programme through the DKCRC (www.desertknowledgecrc.com.au); however, the views expressed herein do not necessarily represent the views of DKCRC or its Participants.

The content has been greatly improved thanks to the major efforts of many reviewers: Steve Morton, Stuart Pearson, Mark Moran, Michael Cooke, Ted Lefroy, John Manger and, in particular, Jan Ferguson, Ian Watson, Mary Stafford Smith and Charlie Veron. The first complete review came fearlessly, and with great insight, from my son David Caffery. Christine Bruderlin redrew the figures in consistent form. My wife Jo Caffery and mother Jean Stafford Smith patiently proofread it all.

In fact, throughout the long journey, my family, Jo, David and James, not only tolerated my distraction but actively inspired the book's completion. I love you all.

Last, the genesis of this book, and the initial ideas expressed in it, came from me, and so these acknowledgements are written under my hand. However, the contents and insights of the book are truly a partnership with my co-author Julian Cribb, whose expressive guidance, encyclopaedic mind, curious English schooling and sympathy for the cause of deserts I most gratefully acknowledge.

Mark Stafford Smith, March 2009.

PROLOGUE

The ancient rocks of the Finke River gorge glow red-gold in the sunrise; the winding valley echoes with the chatter of parrots and the chirps of honeyeaters. Frogs croak from reeds by the still waters in which desert rainbowfish swim; three pelicans coast in, landing with a splash. Bees hum around the whitish gum flowers; brilliant yellow mulga and red holly-leafed grevillea bloom up above the flood plain. The dawn orchestra of desert Australia rises in glorious crescendo.

This scene has been repeated many mornings for thousands of years on the banks of the aptly named Boggy Hole in central Australia. Camping there today, you may overhear a visitor with a cup of tea in their hands, laughing over the noise, 'And I thought this was a desert!'

And therein lies the story of this book.

Waterholes like this one make up a tiny proportion of the desert outback of Australia. Nonetheless, across the whole five and a half million square kilometres, many plants, animals and humans are living with the desert environment. Not merely living, but flourishing in a land that seems hard and harsh to the outsider's eye, yet can be beneficent to those who know how to handle it.

The Australian deserts are reliable only in their unpredictability. Often they are lean in resources, but in some places and at certain times astonishingly rich. They teach their animal and human inhabitants about relationships and networks. They encourage new ideas and the evolution of new strategies and skills. Since European settlement and nationhood, however, Australians have mostly treated the deserts as the backyard of each state, which were protected more by benign neglect than care as the forces of development have blown across the land.

In the past decade, a new alliance has arisen among Aboriginal[1] and non-Aboriginal pioneers in central Australia, exploring what is known about living better in this desert land, and how this knowledge can be harnessed for a safer, more inspiring future. They form the 'desert knowledge' movement.

Desert knowledge is not solely for the desert itself, important as that is. Increasingly, as the forces of climate change, population growth and resource scarcity bear down upon humanity, other regions that we do not think of as deserts are becoming drier and less certain. Even our great cities and fertile plains are increasingly short of water, or beset by economic uncertainty. The wisdom of desert knowledge has a far wider significance than Australians first appreciated. Indeed, it may offer a roadmap to survival in the twenty-first century, not only in Australia, but for much of humanity.

We aim to tell the story of deserts and desert knowledge, and what it all may mean to the world.

* * *

This book is built around two simple but profound and interwoven themes. Firstly, Australian deserts are extraordinary in all senses of the word, and understanding and working with how they function is vital to preserving their environments, their peoples, their wealth and their settlements for the future. Scarce and unpredictable resources result in sparse and remote populations, with consequences for how plants, animals, and humans live.

Secondly, the massive planetary trends being driven by the forces of globalisation, population shifts and climate change mean that other areas of Australia and the world are experiencing conditions that increasingly parallel those of desert Australia. As a consequence, learning how to live well in the desert not only draws up a blueprint for Australia's red lands, but also offers crucial ideas for future living in other regions of our planet that face drying times.

Our first four chapters seek to inspire you with the fascinating features of Australian deserts and their inhabitants; of the fact that, as so many newcomers to the desert realise after a few years, these places really do operate in different ways from less remote regions. The remaining chapters explore various aspects of the livelihoods, settlements, services and governance of desert Australia that must be understood to govern these lands better, and that have significance for the rest of the world.

1

Dramatic deserts

'I collected a great number of most beautiful flowers, which grow in profusion in this otherwise desolate glen. I was literally surrounded by fair flowers of every changing hue. Why Nature should scatter such floral gems upon such a stony sterile region it is difficult to understand, but such a variety of lovely flowers of every kind and colour I had never met with previously. Nature at times, indeed, delights in contrasts…'

GILES 1889[2]

On Monday 2 September 1872, explorer Ernest Giles and two companions were camped on the Finke River in central Australia, near where Hermannsburg would later be established. There had been winter rains that year, and on this rest day he was exploring the vicinity of their camp – an area through which every modern tourist visiting Palm Valley must pass. He made the remarks above in his narrative *Australia Twice Traversed – the Romance of Exploration*, and his question frames the magic and distinctiveness of desert Australia. Why, he asks, was there so much diversity in regions that appeared desolate through the prism of European preconceptions? Asked a different way, what does this tell us both about the desert and about how it differs from the places in which those preconceptions were born? To answer these questions, we must seek to understand that vast outback area of Australia outlined in Box 1.

1.1 Extremes and surprises

The Australian desert is a land of drought and sudden flood, of blowing dust punctuated by carpets of wildflowers, of unexpected waterholes in the midst of dry creek beds, and of the isolated comforts of towns such as Alice Springs embedded in a thousand kilometres of apparently empty landscape.

It is not just the rainfall that varies. Living in Alice Springs, you swelter in summer, unable to recall what 'cold' was ever like. Less than 6 months later, winter chills you to the bone, leaving you incredulous that anything ever felt 'hot'. Driving through swathes of arid woodland, you marvel at the irony that it has been stripped of its leaves by a ferocious hailstorm. A few kilometres down the road, you are forced to flick on the air conditioner in a puny effort to counter the burning power of the sun.

This is the quintessential desert: a land of extremes and paradoxes – variability is at the heart of it all. Every surviving desert organism has come to terms with variability, whether it is an ephemeral plant using the wet pulses and hiding as seeds in dry times, or a desert oak investing in deep roots to find a reliable watertable, a beetle hiding under a stone during the heat of the day, or a nomadic duck flying hundreds of kilometres to find the next

BOX 1: THE DESERT REGIONS OF AUSTRALIA

We use the term 'desert Australia' to mean the arid and semi-arid areas of Australia that are remote from the main population centres. These are characterised by low and erratic rainfall (see Chapter 2) and remoteness, although we shall see that these features lead directly to other important drivers (see Chapter 3).

The desert region spreads across six states and territories of Australia, defining the inland of the continent and even touching the coast in the south and west (left map). At about 5.5 million square kilometres, it covers nearly three-quarters of the country's land area. It is home to about 600 000 people (only 3 per cent of the total Australian population), of whom about 93 000 are of Aboriginal descent.[3]

It is also home to diverse vegetation types (right map[4]), from desert grasslands to low saltbush shrublands and widespread acacia shrublands and woodlands, many of which support species found nowhere else on earth. These reflect the underlying soils (mostly nutrient poor, though the tussock grasslands and chenopod shrubs lands occur on richer clays), and the climate, which varies from winter rainfall dominant in the south to summer dominant in the north.

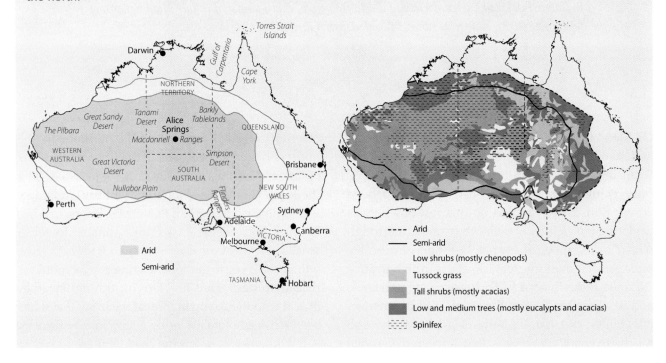

waterhole. Aboriginal people did the same for thousands of years, and today's small businesses, settlements and governments have to find equivalent ways to respond, as we shall see through this book.

Along the sandy banks of Junction Waterhole, where the Ellery Creek meets the Finke River, the poached-egg daisies sprout in their millions in sandy soil after a winter rain (Plate 1a). These ephemeral plants blossom in profusion when water

is around: flowering, setting seed and dying, all within a few weeks. Their seeds wait patiently in the parched soil for the next rains – sometimes for weeks, sometimes many years. When the rains arrive, not all the seeds germinate – after all, it may be a false alarm, and over eons the plant has 'learned' not to take the risk of putting all its eggs in one basket. So, the seeds have a variable degree of dormancy – some will germinate the first time, some the second, but botanists have found that a handful will not germinate for decades.

Out on the harsh red plains there are mulga trees, which will live for four or five decades – even centuries if they have their strategy right, and are a little lucky. Acacia shrubs such as mulga dominate nearly a third of the Australian deserts, proof that they are not careless strategists. A long-lived perennial, mulga puts its greatest investment into its root system, which spreads wide and deep to harvest as much sparse soil moisture as it can. It has also developed a canopy shape that funnels the rain, when it does arrive, down its stem and into the root zone. Mulga will survive most dry periods, but every now and again a season comes along that is too severe – as one did in the 1960s when even these tough trees died. Mulga flowers and seeds after most rains, producing seeds that germinate more freely than those of ephemerals, with a good chance that some will survive that terror drought.

Other plants are much more selective in how they use the desert. The river red gum grows beside the waterhole because its strategy is to exploit the permanent water lying below the river floodplain. It uses water freely in this land that seems so dry, though it has worked hard to build a large root system to be able to do so. With this strategy, it can live for more than a century if all goes well. But, like the mulga, it must take out insurance against the day when the rains fail for so long that even the water table below the riverbed dries up. So, the river red gum flowers every year, as regular as clockwork, scattering thousands of potential offspring around the floodplain. If the rains are heavy and floods come, these are dispersed widely.

All desert life is adapted in one way or another to cope with erratic pulses of rainfall and long, tough, dry times between. Some, like the ephemeral daisies, exploit the pulse then hide away. Others, like the perennial mulga, invest the pulse in deep roots and plentiful seedlings, many of which die. Each pulse of life-giving rain is different from the previous one – shorter or longer, heavier or lighter, stormy or soaking – and offers an opportunity for a different survival strategy to shine. The climate of desert Australia really is exceptionally variable on a world scale, so a great diversity of adaptations can co-exist.[5] The extreme challenge this poses to living creatures is the source of the continent's vitality and biological inventiveness.

Rainfall variability affects not only plants but also animals and humans.

Animals can use equivalent strategies to thrive in the deserts. Mouse-sized desert marsupials, such as the tongue-twisting fat-tailed false antechinus, achieve a similar effect to the extensive root system of mulga. These tiny carnivores succeed by having a home range that is large for their size, so they can gather sufficient of their sparse insect food to store as fat in their tails against the bad times (Plate 2c). They can only do this when the food resources are not spread too thinly, otherwise they have to spend more time searching than eating. As a result, in central Australia, they are most common in the mountain ranges, where the insects persist through dry times thanks to small pockets of concentrated plant production in rocky cracks and crevices.[6]

Like the river red gum, other creatures seek out a more moderate and predictable microclimate. For example, by being small, they can find shade and protection in a clump of spinifex grass. Small creatures can also divide up the food resources in many ways; there are often several species of skinks inhabiting the same spinifex clump, but avoiding competition by foraging in different parts of it, or at different times of day or in different seasons of the year. In 1990, herpetologist Craig James found the world's highest local diversity of skink species in a patch of spinifex grasslands 30 kilometres south of Alice Springs. He found up to 11 different species of skinks in an area of 50 square metres (and up to 20 species of lizards of all types); as many as seven species of skink might shelter or forage in the same spinifex hummock over the course of a day.[7]

Animals can be ephemeral, just like plants. Shield shrimps are famous for living for a few days in puddles on the top of Uluru (Ayers Rock), then surviving as a hard-shelled egg that is blown around on the wind until it ends up somewhere where rain comes again. Many insects, such as the desert locust and various mites, lay resistant eggs and wait. Other animals avoid the bad times as adults – for example, burrowing frogs hide half a metre or so under every sandy watercourse and most sand dunes in inland Australia.[8] They pop up as soon as the ground wets enough to alert them to serious rain, which is likely to have created puddles that will last long enough for them to breed. In the middle of the Tanami Desert in 1990, after years of dry weather, desert ecologist Steve Morton was soaked by a downpour, but astounded to find thousands of frogs appearing from the sand plain within hours – they must have been waiting in huge numbers for at least a decade, husbanding their meagre water supply. Both in the Tanami and in South Australia, these frogs have been recorded at densities of 50 to over 200 individuals per hectare, often ten to a hundred times more biomass than that of mammals in the same environments – showing how successful the strategy can be in these environments, which seem so unlikely to support frogs.

Compared with plants, animals have an extra trick up their sleeves: they can move within their lifetimes to seek out new food supplies, sometimes over huge distances. The budgerigar is a classic desert nomad, moving in large flocks from waterhole to waterhole and seed source to seed source (Plate 3c). A few species, such as rainbow bee eaters and woodswallows, may migrate regularly in and out of the arid zone. Still others, such as the pelican, essentially live outside the desert, but come in to exploit good conditions. Many aquatic birds spend much of their time in the south and east but have their most spectacular breeding seasons on the rare occasions when water fills the salt pans and waterholes of the Lake Eyre Basin, as it did in 2009 (Plate 3d).

The main strategies bred by the variability of Australia's deserts – and which we will explore in desert dwelling humans too – are thus to *persist* as active adults (putting down deep roots or adapting in other ways, such as perennial plants, reptiles, resident birds and many mammals) or to *move in time* (avoiding dry times in inactive forms, such as

ephemeral plant seeds, insect eggs, bulbs and aestivating frogs) or to *move in space* (such as nomadic or migratory birds and insects).[9]

In the course of adapting to variability, Australian plants and animals have explored every alternative and combination in time and space – from short-lived to long-lived, and from sedentary to highly mobile. Not only that, they can be flexible – for example, kangaroos know to expand their ranges when it is dry, and may even move substantial distances *en masse* in bad times, although they are not normally nomadic. Humans, of course, are the pinnacle of flexibility, although we do not always take full advantage of this.

Deserts are not only variable, they are also unpredictable. Being unpredictable is different from being variable – regular waves are variable but perfectly predictable. At sea, it is violent, unexpected waves that sink ships and drown fishermen, not just big seas. The uncertainty of when the extreme is going to happen is as important as the uncertainty of how severe it will be.

The problem is more profound than this, though, because combinations of events are even more unpredictable. For example, the 1970s were the wettest years on record in central Australia – but they were also years when rain came more in summer than winter. This combination – with big rains in summer – benefited grasses, which mostly prefer to grow in warm conditions. By the late 1970s, central Australia had huge grass fuel-loads and started to experience many wildfires. The 1980s were less wet, but were dominated by big cool-season rains. These rains benefit forbs and small shrubs such as salt-bushes. The grasses of the 1970s declined and instead there was a decade when the vegetation was full of prickles from winter-growing bindii and copperburr. These do not produce the right fuel for fires, so the 1980s were almost fire-free around Alice Springs and people forgot the risks of careless campfire behaviour. But, in the early 2000s, the summer rains returned again for several years, and around 20 million hectares went up in smoke in the arid zone part of the Northern Territory in the 2002–2003 fire season.[10]

Variability and unpredictability in the deserts are not about water alone but also about its interactions with other factors such as temperature, fire and fertility and, increasingly, human activity.

For example, although Australia is hot, we do not usually regard it as being as extreme as the middle of the Sahara or the Rub'al Khali of Arabia. Yet, in 2003, a satellite awarded a 25 square kilometre dot of Mitchell grasslands north-east of Winton in Queensland the dubious honour of being that year's hottest pixel of land surface on Earth, at 69.3°C.[11] Australia is the land of extremes and surprises in all sorts of ways.

The wild variability of rainfall or temperature is not the only surprise. The timing and combination of extreme conditions – all of them unique and unlikely to recur in our lifetimes – are what make this land fundamentally unpredictable. We have to think about living in and managing such lands very differently from a neat and regular European landscape. But variability is not the only special feature of desert Australia.

1.2 Sparseness and space

The Australian desert lies on the old, worn-down bedrock of an ancient continental plate. Parts of Australia date back at least 3.5 billion years and contain fragments 4.4 billion years old – close to the origin of the planet itself. Then there are areas of hills south of Tennant Creek (in the Davenport Ranges) that have been continuously exposed to the stars for over 500 million years and are among the oldest known landscapes on Earth.[12] In half a billion years, every gram of nutrients has been washed off, blown away, and re-sorted countless times. Not all of desert Australia is quite this ancient – for example, today's sandy landforms are only a few tens or hundreds of thousands of years old – but most of the soils are made of recycled, ancient and impoverished materials. The sands and loams that cover two-thirds of the desert are low in phosphorus and nitrogen and other elements needed to sustain life, which have been leached from the mother soils by eons of erosion and borne

away on wind or flood. Not only are these areas limited by water most of the time, but, even when water is plentiful, plants and animals are often limited by nutrients.

As a result, most Australian desert landscapes are unproductive and so cannot sustain large forests or animal populations. They are extremely challenging for agriculture, so only a sparse human population inhabits the outback.

Where nutrients are scarce, they are also immensely precious. Even small concentrations, such as in a little flood-out from a creek, can greatly magnify the amount of nutrients in a local patch of landscape. Thus small, almost imperceptible, patterns of movement by water and wind on the landscape are vital for creating new opportunities and niches for plants and animals that have developed ways to seek them out.

On a much larger scale, huge water movements down rivers such as Cooper Creek result in sunny floodplains that, for a brief period after flooding, are as productive as the richest agricultural systems or a tropical rainforest, fixing carbon at a rate of 2 to 5 tonnes per hectare per year.[13] Queensland freshwater ecologist Stuart Bunn has measured growth rates of organisms that are so high that one day's growth on the vast inundated floodplain was equivalent to 82 years of production in the waterholes during dry periods. Such flooding occurs only for a month or two every few years (Plate 1d), but the resulting burst of productivity supports waterbird breeding on a massive scale that could not occur if the production was spread evenly over those years.

Deserts, in other words, are not only places of general scarcity – but also places of immense and usually very transient local richness.

Australian deserts vary in space as much as they vary in time. Broad tracts are very infertile most of the time, but patches can be incongruously rich. People who want to live on these lands have to think about using and managing for the extremes, just as the floodplain fish and waterbirds do. This is a very different paradigm from that of European-style agriculture, which depends on consistently fertile soils and reliable rainfall. The plants and animals of

Australia's deserts have had millions of years to work out how to take advantage of these variations – and newcomers who have lived here only a few decades can learn much from them.

Sparse resources mean that population densities cannot be too high. Plant biomass in arid lands typically peaks at a few tonnes per hectare above ground, whereas temperate forests average 150 tonnes per hectare. But individual species have interesting stories too – some huddle together in rich patches, hanging on to whatever local benefit they have found. Examples are a whole community of plant and animal species that hide in moist gullies in the arid mountain ranges, such as the cycad, ferns and the well-known Palm Valley palms in the Krichauff Ranges of central Australia. Others are extraordinarily isolated. An amazing example is the inland worm, three populations of which were found by the Horn Scientific Expedition[14] in 1894; they were then not seen again until a fourth population was found at Uluru in the 1990s. Worms are not the most iconic of organisms, and seem unlikely to cause a surge in tourism, but it is a matter for great wonder as to how these moisture-dependent creatures ever spread across arid catchments, rocky hills and sand dunes to these four unconnected sites, hundreds of kilometres apart. Have they simply survived in these tiny, fragile pockets for 7000 years since the climate was much wetter? They have been seen so rarely that the answer is not yet known to science.

The curious fact about sparse production is that, while there are few nutrients per square metre to sustain large populations close together, there are a *lot* of square metres in desert Australia – about 5 000 000 000 000 of them, in fact. So, if you are good at collecting a small amount of nutrients from a very large area, there is plenty to be had out there.

One good strategy for doing this is to be small, numerous and well-organised – the job description of termite and ant colonies. Both of these insect groups use their colonies to forage over large areas, collecting sparse resources. Although ants are not averse to collecting other animals (including termites!), termites specialise in breaking down dead plant material – which is plentiful, but of the very poorest quality nutritionally. Different species cut and collect dead grass, or chew up dead wood, and cart it back to their nests. There, they process it with the aid of specialised gut microbes that break down the cellulose into something the termites can digest. This is hard work, and termites must be slow and steady workers. As a consequence, they are rapidly out-competed by other animals in richer environments, but in areas with low nutrients such as the spinifex sandplains they are without peer (Plate 2d).

Plants and animals that thrive in the deserts are those that use the pulses in time, or richer patches in space, or simply harvest sparse resources efficiently. Humans in deserts must obey the same laws of survival. There are necessarily fewer people in the desert (Australia's overall population density averages 2.5 people to the square kilometre, but only a tenth of a person per square kilometre in the outback). Thus, as the early settlers found, the fourth challenge of living in a desert land is distance.

1.3 Remoteness and relationships

If you are a long way away from everyone else, you either have to be very self-sufficient or you have to be very good at communication and travel.

An unexpected illustration of this is the bloodwood apple gall.[15] This gall forms only on the bloodwood tree, which is widely scattered in open woodlands in central Australia. The gall is formed by a tiny female wasp, which injects chemicals into the plant to trigger the growth of tissue around her. She feeds on the sap as she develops. After a year or so, she lays a series of eggs that all hatch into tiny male wasps; these feed on the inside of the gall and mature, although they are still very small. Just as her sons are close to developing their wings, she lays a second batch of eggs that are all female, which hatch and climb on to the tails of their brothers. The mother then opens the plug that seals the gall (she had stuck her abdomen into a hole while the gall grew for the previous year or so, for just this purpose), releasing them into the night air at a propitious time, and then she dies. The males fly around to seek another bloodwood where there might be an adult female to mate with. Eventually, the females disembark from the survivors and the whole cycle starts again. The main role of the males seems to be to provide transport for their sisters between widely dispersed host trees.

At a much larger scale, ornithologists have been using satellite tracking to gain insights into the extraordinary movements of the desert's ultimate nomads: waterbirds. Ornithologist David Roshier has tracked birds that may move hundreds of kilometres from one water body to another to find transient specks of water as small as a farm dam. He observes that it is the ability to use a shifting mosaic of wetlands – which come and go at different times in different regions from the Murray–Darling to central Australia – that sustains such a large population of waterbirds in Australia.[16] All this depends on the birds' ability to move over great distances, successfully searching for water-bodies on a continental scale.

Tens of thousands of years before any scientist analysed the need for mobility, Aboriginal people too were moving across the arid Australian landscape, taking advantage of its areas of richness. Their fantastic oral histories – dreamtime songlines that knitted relationships and trade across the continent – show how intensely this strategy was developed over millennia of living within the desert landscapes (see Chapter 4). It was a lesson not lost on some early pastoralists, who picked up clues from the expertise of their Aboriginal stockmen. Although Australian land tenure systems have mostly promoted a sedentary way of life, from Sir Sidney Kidman onwards, pastoralists have been finding ways to move stock around the country. At first this involved droving animals on foot down the 'long paddock' and trails such as the Canning Stock Route, but more recently by cattle trucks – the outback 'road train'. Whether by owning leases in places with different climates and conditions, or negotiating stock grazing arrangements with other pastoralists, mobility has long been a strategy for the grazing industry.

Living in remote places has several more impacts that are important to our story. Firstly, it means that desert dwellers are a long way away from the main

concentrations of people, where the markets are, where things are made, services are provided and most of the rules are developed. So, they are remote from people who might want to buy their beef, shape their gold, hang their art or even have their offspring. It calls for smart ideas for staying in touch and keeping networks up to date. For example, Yuendemu's remote art centre in the middle of the Tanami – a dusty 300 kilometre drive from Alice Springs – would fail if it could not market directly over the internet to purchasers as far away as New York. Desert people have always had to invest heavily in knowing their market value chains – a bit like waterbirds knowing where to move to. Aboriginal peoples traded ochre for painting and nicotine-containing pituri leaves across the continent from Western Australia to Queensland – far beyond the distance that individuals would normally travel. We'll revisit the opportunities this opens up for desert businesses in Chapter 6.

Secondly, it means that desert dwellers are a long way from the centres of political power. Outback problems and issues seem distant and trivial in Australia's great coastal capitals, even if the outback occupies three-quarters of the nation, generates a sizeable share of its export income and furnishes its iconic self-images. This breeds a special culture of self-reliance, but, at the same time, it may lead to frustration with how the rest of the world operates. We'll look at the implications of these issues for settlements in Chapter 7 and governance in Chapter 9.

Thirdly, it requires special social technologies. We tend to think of Aboriginal traditional knowledge as being about bush foods and medicines, and indeed there is plenty of this. But there are deeper and more significant layers of such knowledge. Baldwin Spencer, writing insightfully on his first visit to central Australia as part of the Horn Expedition in 1894, put it this way:[17]

'To the rules of the community [they] conform quite as strictly, in fact perhaps more so than the average white man does to the code of morality which he is taught.'

These rules encompassed a diversity of arrangements that are designed to help people survive in harsh environments, and to ensure that the resources on which they depended were looked after. It included intensive networks of reciprocal obligations, which meant that when conditions were tough in one region, people could borrow space from their neighbours for a while. It also created sophisticated ways of splitting the responsibility for looking after totemic plants and animals to ensure that knowledge and management would not be compromised through accidental death or a failure to act, as we'll see in Chapter 4. In their own ways, the frontier white settlers found that they needed close support networks too, which persist in the mythologised self-reliance and strong social bonds of the outback today. Throughout the book, we shall explore the lessons to be learned from both Aboriginal and non-Aboriginal culture in desert Australia, as well as the growing alliances between them.

1.4 Lands of beauty and bounty

Despite the incredible variability in water supply faced by desert plants, despite the sparse and patchy nutrients – despite everything – desert Australia is more productive than anyone would expect. This is because so many organisms, particularly the persistent perennials, find ways of lessening the variability by their presence. The classic example is the river red gum (Plate 4a). This majestic individual has its feet in water, and it is certainly not short of sunlight, so it can photosynthesise to its heart's content. It produces leaves, flowers and surplus sugars, which are eaten by a whole suite of insects, birds, and mammals. These, in turn, are preyed upon by another layer of consumers – more insects and spiders, lizards, birds of prey and humans.

In drier environments, mulga fulfils the same role: supporting many lerps (small sap-sucking bugs), which are tended by ants, which themselves support lizards. In the most infertile sandplain environments, spinifex persistently produces dead material for termites. The termites store it and process it with the help of their symbiotic gut bacteria. They become food for predators, such as the king crickets and lizards (Plate 4b), which are in turn eaten by birds, mammals and humans.

Humans cope with deserts in similar ways – by seeking either to stabilise the extremes or to capitalise upon them. For example, in 1991 rangeland scientist Barney Foran analysed two neighbouring cattle properties in central Australia that received the same rainfall, but which were operated quite differently.[18] One property's strategy was to stock lightly, but to get excellent growth rates and good sale values; this property had a low, but steady, income year to year – which was much *less* variable than the climate input. The other bought and sold stock to use the grass in good years and avoid cattle deaths in bad years; this property had wildly fluctuating cash flows – which were much *more* variable than the climate. Both seemed to make a similar long-term average cash flow, but, while one had dampened down the climate signal, the other had greatly exaggerated it. Foran called one 'easy living' and the other the 'heart attack' strategy!

In the face of adversity, plants, animals and humans in deserts have all developed strategies not just to survive but to thrive – and to do so by taming or exploiting the variability and resource limitations thrown at them by nature.

Deserts are associated with desolate wastelands, as expressed by a frustrated John McDouall Stuart when he probed into an endless rolling sea of mallee on the Eyre Peninsula in August 1858: 'Today's few miles have been through the same *dreary, dreadful, dismal desert* of heavy sand hills and spinifex with mallee very dense, scarcely a mouthful for the horses to eat. When will it have an end?'[19] (He provided the emphasis!)

It was tough going for Stuart's horses and his men that day, so one cannot blame him for what he wrote tiredly in his diary. However, the widely held perception of deserts as barren and worthless is ill-founded at many levels. Although the environment is harsh, used carefully, it can still support domestic grazing in some regions. There are pockets of particularly spectacular scenery that attract tourists, such as waterholes in the Flinders Ranges and monoliths such as Uluṟu (Ayers Rock) and Mount

Augustus. Other tourists are attracted by resources that are not so patchy – the simple reality of open spaces and skies filled with stars down to the horizon. Rich deposits of iron ore, bauxite, natural gas, oil, opals, tin, gold and silver speckle the interior, permitting intensive local exploitation. Sunlight, on the other hand, is a universal energy source in deserts, which Australia enjoys more than any other country on Earth. Then there are small areas of intensive irrigated horticulture, such as on the Ti-Tree Basin north of Alice Springs.

The desert may be dry but, per head of population, it is a rich and most productive region. It generates over $90.5 billion a year in gross revenue for Australia, mainly from minerals, energy, pastoralism, cultural and eco-tourism, and art.[20] To put this in context, gross domestic product for Australia was about $1233 billion in 2007/08. Thus, the desert produces over 7 per cent of the nation's domestic product from 3 per cent of the population, and most of this product is real and solid – not virtual financial exchanges. On average, each desert resident contributes around $80 000 in goods and services to the Australian economy each year; this is half as much again as the national average. A disproportionate part of this production goes to export, so that the desert contributed a significant share of Australia's total commodity export earnings of around $150 billion in 2007/08.[21]

All in all, desert Australia is a surprisingly bountiful contributor to the nation. It is valuable for many other reasons. It is the home of a massive natural heritage of species that exist nowhere else on Earth. Aboriginal culture persists in the desert and other remote regions of Australia more than anywhere else. Images of the outback pervade Australian media, advertising, poetry, music, art and language. Although some people find the desert harsh, it is also a great place to live – a recent report found that remote Australia tops the nation on every measure of social capital. People in remote Australia are more likely to feel part of the community, get together at least weekly with friends and relatives, volunteer more each week, belong to community-based, hobby or sporting organisations, easily able to raise $2000 in one week

in an emergency and can usually find someone to help out when needed.[22]

Sustaining the desert matters for economic and emotional reasons, as well as for the lessons that the desert can teach.

1.5 A dry heritage
Worldwide, the drylands and deserts are also of profound significance to our human culture and development – and so to our common future.

Indeed, humans may be the children of drought. Our ancestors first separated and began to evolve away from the other apes during a period of extreme aridity 6–4 million years ago – a time known as the Great Messinian Salinity Crisis. This was a period in which the Mediterranean sea dried out completely and deserts around it expanded. In Africa, the forests shrank and grasslands spread, creating intense stresses for all animals, including human ancestors. These scarcities compelled early hominids to choose between a life in the dwindling forests and one out on the growing savannah – and a choice between eating fruit and eating meat. Our ancestors were among those to opt for the grasslands – and strong social bonds, an upright posture, the ability to communicate, the use of tools and fire and a rapidly enlarging brain all helped survival in this challenging environment.[23]

The first primitive tools were shaped by early hominids surviving in the arid lands of Ethiopia and East Africa around 2.2 million years ago. By 1.4 million years ago, our relatives in southern Africa were using fire to defend themselves from predators, and possibly to cook meat. By a million years ago, they were fully upright, their brain was expanding rapidly in size, and their use of tools had grown spectacularly, both in range and utility. By 130 000 years ago, hominids were, for all intents and purposes, fully modern in physique, and it is these people that most palaeontologists consider set out from Africa to inhabit most of the Earth over the ensuing 100 000 years, replacing or out-breeding all other pre-human populations in the process.

Many of the early great civilisations, such as those of the Tigris–Euphrates, Indus, Nile and Yellow

River valleys, arose where water was plentiful in regions surrounded by dry landscapes. These environments contributed many of the most readily domesticated grasses and herd animals. Agriculture, the keystone of civilisation, and irrigation, the keystone of city building, are both based on grain production, which originates in the grasslands. Without the continual stresses and pressures that the drylands have exerted on their inhabitants, most humans might have been content to remain hunter gatherers to this very day. Thanks to these continual and variable pressures, the drylands are a fountainhead of innovative ideas for survival.

Over history they have seen the birth of farming, irrigation and cities, which, in turn, introduced pottery, textiles, metalworking, writing, architecture and building, and social, legal, political and philosophical religious systems, many of which persist to this day. Sciences such as mathematics and astronomy were born among desert peoples. Religions such as Judaism, Christianity, Islam and Zoroastrianism had their genesis in the drylands. It is probable the first explosives came from the nitre to be found in deserts and the earliest steel was made 1000 years ago in the city of Merv in Central Asia, giving the means of victory to would-be conquerors.

In short, many important innovations that have influenced human development are themselves the product of a dry landscape whose variability confronts its inhabitants with continual survival challenges: forcing them to use their brains, creativity and originality, and often to borrow ideas from nature.

Throughout time, deserts have tested the limits of human, animal and plant ingenuity and resourcefulness in survival. For those who look and take the time to understand them, they offer opportunities, challenges and advantages unlike other regions. In an increasingly variable and unpredictable world, they hold secrets not only for local survival, but for planetary prosperity.

2

Spreading deserts: past and future

'So I'll see you out on the mulga and spinifex plain / Anytime Tjilpi I'll be coming back this way / See you out on the mulga and spinifex plain / Just light me a fire and I'll be home again'

WARUMPI BAND, NEIL MURRAY[24]

The deserts of the world are our human heritage: dynamic, challenging, profuse and, above all, different. They have been a home for humans since we first evolved. Managed badly and degraded, they can also be hostile, barren and destructive, as many have found out.

Throughout pre-history, deserts have affected humanity, both in Australia and around the world. Today, a third of the world's land surface is semi-arid or drier – 'desert' as we use the term here (Box 2) – and a fifth of the world's population lives there. These deserts are on the march again, with profound implications for humanity. By mid-century, more than two in every five of the Earth's citizens is likely to be living in a desert or under desert-like conditions. For governments and international agencies alike, desert issues will be of paramount significance; unless these are adequately resolved, the deserts may become a fount of political, social and economic turbulence and a source of instability for humanity at large.

Desert Australia presently occupies three-quarters of the continent's land area (see Box 1). Its 600 000 people live in four small cities, 860 remote Aboriginal communities, 7000 pastoral stations and earn more export income per resident than any other part of Australia. This desert region is also about to expand due to the warming and drying of the climate and to land degradation, although the rate and extent of this expansion is not yet clear.

As the trend intensifies, cities such as Canberra, Dubbo, Goulburn, Horsham, Merredin and even Toowoomba may one day find themselves with desert-like rainfall and evaporation regimes, in landscapes resembling what is known as pastoral country today. Well over a million Australians will be desert dwellers and the nation will have a great many more towns like Alice Springs. These trends and their impacts have a long history and will create a challenging future.

2.1 A drying history

Australia has been generally drying since it parted company with Antarctica around 80 million years ago and began its geological gallop – five centimetres or so a year – towards Asia. As it forged its way into the mid-latitudes, the continent fell under the

BOX 2: THE DESERTS AND DRYLANDS OF THE WORLD

Deserts are defined by their low rainfall, lack of moisture and high rates of evaporation; their dominant ecosystems range from ultra-arid deserts to grasslands and open woodlands, such as those that cover much of Australia. The world's drylands are classified as hyper-arid, arid, semi-arid and sub-humid, in order of decreasing aridity. In this book, we talk of the hyper-arid, arid and semi-arid lands as 'desert', although Australia does not contain any hyper-arid lands like those of the central Sahara in Africa or Atacama desert of Chile. Both of these include regions with less than 25 millimetres long-term average annual rainfall, whereas the lowest long-term annual rainfall in Australia today is experienced around Lake Eyre at about 100 millimetres per year.

Thus the term 'drylands' encompasses four types of arid lands, and we use 'deserts' to mean the three drier types, particularly where they are remote from population centres. Many of the issues raised in this book are relevant to all drylands, but often less critically to the dry sub-humid regions, or to the few desert regions that are densely populated. The global distribution of dry regions is shown below.[25]

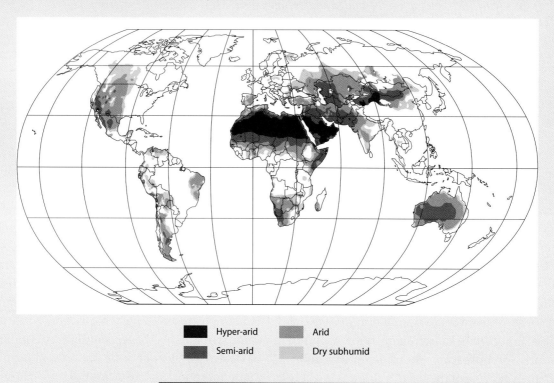

■ Hyper-arid	▨ Arid
▨ Semi-arid	▨ Dry subhumid

| Type | Aridity index | Current area | | Population (2000) | |
		Area (millions sq. km)	Per cent of global land area	Total (millions)	Per cent of global population
Hyper-arid	<0.05	9.8	6.6	101.3	1.7
Arid	0.05–0.20	15.7	10.6	242.8	4.1
Semi-arid	0.20–0.50	22.6	15.2	855.3	14.4
Desert total	**<0.50**	**48.1**	**32.5**	**1199.4**	**20.2**
Dry subhumid	0.50–0.65	12.8	8.7	910.0	15.3
Drylands total	**<0.65**	**60.9**	**41.3**	**2109.4**	**35.5**
World total		147.6	100.0	5942.0	100.0

BOX 3: WHY DESERTS ARE WHERE THEY ARE

Spin the traditional globe on its axis and you will see a yellow blur of arid country ringing the Earth above and below the equator (see Box 2). These are the world's great desert belts – around 25 degrees of latitude into each hemisphere – the Sahara and Arabian deserts, the Gobi and the Chihuahuan in the northern hemisphere, and the Atacama, Kalahari and Australian deserts in the southern.

These desert belts are caused by the behaviour of the atmosphere in what are known as Hadley cells. Heated by the sun, warm air made moist by evaporation from the oceans rises over the hot equatorial belt. As it rises and cools, it drops its moisture in tropical downpours, then corkscrews outwards from the equator delivering cold dry air to the mid-latitudes, or subtropics. In this region, it sinks and warms up, becoming very dry and hot by the time it reaches the ground and blows back towards the equator to close the loop of the Hadley cell.

The result is a hot region with low and erratic rainfall where most of the warm deserts of the world occur. There are other deserts that occur in rain shadows where ocean currents or uplifts over mountains cause the air to be dry, such as the coastal Namib desert and the continental Tamaklikan desert of western China. Equally, in other regions – particularly the east coasts of continents – the natural tendency for the Hadley cell circulation to create dry conditions is offset by coastal moisture blowing inland.

In summer, the Hadley cell belts of dry air move polewards with the tilt of the Earth. In the case of northern Australia, this allows monsoonal rains to penetrate further south. In winter, the belts move north, and rainfall from the southern oceans extends further north into southern Australia. However, with global warming, the Hadley cells will get larger, pushing the dry belt permanently further away from the equator. This will bring dry times to many of the world's key food baskets, such as the American Midwest, the Indo-Gangetic plain in India and Bangladesh, and Australia's great southern wheat belt.

influence of the dry climatic band that girdles the Earth. Deserts are common at this latitude because of the behaviour of huge atmospheric circulation patterns called Hadley cells (see Box 3). Australia currently straddles the Tropic of Capricorn – 23.5 degrees south – which, in turn, bisects three of the Earth's great arid regions: the Kalahari, Atacama and central Australian deserts.[26]

As it has moved slowly north, the Australian continent has undergone periods of extreme aridity and significant wetness, creating enormous variability within the general drying trend. This variability has continued into recent times. The continent has not moved far in the past 135 000 years, yet, even during this period, Lake Eyre – Australia's greatest 'rain gauge', which fills or not according to rainfall over inland Queensland – has been both 25 metres deep and bone-dry at different times.

Reflecting these enormous variations, trees and sand-dunes have ebbed and flowed across the landscape in a way that is hard for the continent's present-day inhabitants, who are used to a comparatively wet phase, to imagine. For instance, at the peak of aridity – a mere 21 000 years ago – much of the continent was treeless, except for the coastal fringe and a few valleys; and sand-dunes blew from the inland to the sea on all but the eastern coastline. Yet, 25 000 years earlier, Aboriginal people had been drawn to the continent's heart by the extraordinary wealth of its lakes and their fish, bird, animal, shellfish and plant life. Places such as Willandra Lakes in New South Wales, Lake Carpentaria (today the marine Gulf of Carpentaria) and the Amadeus Basin in central Australia were freshwater paradises, of which few traces now remain.

All in all, the last million years have been a time of profound disturbance in the Earth's climate, with no fewer than six ice ages occurring in the northern hemisphere, with proportional impacts felt in the south. Each ice age has coincided with a drying phase in Australia, and each interglacial (the period between two ice ages) with a warmer, wetter climate and high sea levels like those of today.

An Australian time traveller who arrived 120 000 years ago – in the warm phase between the last two ice ages – would find the landscape astonishingly wet compared with today. Lake Eyre brimmed with water, the Gulf of Carpentaria was an enormous inland freshwater lake, while the Amadeus Basin cradled a chain of huge shallow lakes stretching from central Western Australia to south of Alice Springs. The Murray-Darling Basin was dotted with water-bodies and wetlands that have long since vanished. (The early European explorers of Australia were right – there was indeed a fabled inland 'sea'. They simply arrived 120 000 years too late to sail it.)

Fast-forwarding in time to 50 000 years ago, our traveller would find the landscape still relatively wet and well-vegetated, with large water-bodies. This was the period in which Aboriginal Australians first spread extensively across the inland.

However, vast changes were afoot. The monsoons that replenished the great lakes of the Centre from the north began to fail. Wildfires seared the landscape and hastened its drying out. Combined with pressure from human hunting and burning activities, these changes brought huge ecological stresses for Australia's wildlife, causing the extinction of the giant marsupials and birds around 46 000 years ago.[27] By 36 000 years ago, a new drying trend was firmly established as the most recent ice age took hold.

Reaching the peak of this ice age, around 21 000 years ago, our time traveller would witness the most arid conditions in the recent story of Australia. Vast fields of sand dunes rolled their way to the northern, western and southern coastlines, and even to the east coast of Tasmania. Lakes vanished. The sediments of playa lakes – the last salty vestiges of the 'inland sea' – collapsed and blew away as they dried out, creating an epoch of fierce dust storms

that left their deposits as far afield as Antarctica. Woodlands and forests died back, the remnant trees clinging to tiny valleys and refuges or retreating to the coastal fringes. Spinifex and grasslands expanded massively. With the loss of vegetation, salty groundwaters rose and salinity claimed large tracts of the landscape. The Great Barrier Reef became dry land, with Aboriginal people inhabiting its caves. Glaciers formed in Tasmania and on the mainland's Eastern Highlands.

Moving forward in short hops through time, our traveller would find that even within this hyper-arid period the Australian climate was nevertheless wildly erratic and there was a succession of wet and dry pulses. If the Australian deserts can be imagined as a sea, their arid tides have flowed and ebbed across the face of the continent no fewer than seven times in the period from 30 to 13 000 years ago.

The closing phase of the latest ice age, 15 000 to 9000 years ago, would show our time traveller further dramatic change. The climate warmed and became much wetter again, woodlands migrated back into the dry country and eucalypt forests expanded inland and into the uplands. With the melting of ice previously locked in the polar ice caps, sea levels rose by 150 metres. This severed the continental mainland from Tasmania in the south and from Papua New Guinea in the north, inundating Lake Carpentaria and bequeathing the continent its present familiar shape. Rainforests spread. Perennial grasses, which dominated the deserts throughout the dry period, retreated before this invasion. The northern 'wet season' pattern became established.

This bountiful period lasted until around 6500 years ago when the time traveller would notice signs that the continent is drying out once more, entering what the Australian National University's Patrick de Deckker terms a 'hydrological deficit'. Sediments from lakes in south-eastern Australia show changes in wind speeds and rainfall that suggest that, until around 6500 years ago, El Niños (which bring drought conditions) were virtually non-existent in Australia and La Niñas (which bring wetter conditions) ruled the continental climate.[28] Then, quite suddenly, the climate seemed to flip and, for the last 6000 years or so, El Niños began to dominate. In the last 600 years, there has been a further increase in aridity, which has been accelerated by European settlement in the most recent two centuries. Land clearing has amplified the hydrological deficit by reducing transpiration of water by vegetation, raising saline water tables, increasing erosion and dust storms, producing hotter conditions and, it is now thought, contributing to declines in local rainfall.[29]

A challenge faced by Australians in comprehending such vast swings between dryness and wetness, desert and forest, is that our 'snapshot' of the climate – based on weather observations – is little more than 100 years long. This is barely 1 per cent of the time since the last ice age and one millionth of the time since the continent itself sheared away from Antarctica. Though our experience encompasses droughts and floods, it does not reveal the extreme swings that have occurred even in the recent geological past.[30] In that sense, De Deckker fears we may still hold a naively optimistic view of how wet Australia is. The fact that global warming will trigger more dry times makes this optimistic view even more misleading.

2.2 Today's variable, unpredictable and extreme climate

Four-fifths of the Australian land mass receives fewer than 600 millimetres of rain a year on average, and half of it receives fewer than 300 millimetres. However, these figures conceal the extreme variation that is a particular feature of Australian deserts. Records for the town of Alice Springs, for example, show years with an astonishing high of 903 millimetres and an equally remarkable low of 54 millimetres of rainfall.

When we think of deserts, we think first of their dry climate – and this, it is true, is their most important environmental driver. However, although all deserts are dry on average, they differ in a second key feature – the variability and unpredictability of their water supply. Deserts caused by ocean currents, such as the Namib in south-west Africa and the Atacama in Chile, are dry but relatively predictable (rainfall in the Atacama is predictably zero!). Indeed,

their plants obtain much of their moisture from regular fogs. But the great mid-latitude continental deserts, such as the Sahara, Gobi, Kalahari and the Australian outback, all share high variability in rainfall from year to year. In seminal research, Australian climatologist Neville Nicholls[31] showed that, as a rule, the variability of annual rainfall increases with decreasing total rainfall and closeness to the equator. In the subtropics, these factors combine to yield the highest variability in precisely the zone where the great deserts are found (see Box 3).

However, Australian deserts – unlike the Sahara or Gobi – are subject to a critical additional factor that affects rainfall variability: El Niño. El Niño and its counterpart, La Niña, are the weather patterns caused by changing sea surface temperatures in the Pacific Ocean, which then affect rainfall over Australia (and over western South America, although in an opposing way: when eastern Australia is dry, Chile is wet). Australia's drylands experience the usual desert variability both within years and between them. But overlaid on these fluctuations are substantial wet and dry periods associated with El Niño and La Niña, which occur on average (though not regularly) every 4 years.[32] Furthermore, there is now good evidence that the severity of El Niño events is altered, in turn, by another cycle of about 19 years, the so-called Inter-decadal Pacific Oscillation.[33] We do not yet understand what drives this oscillation, though some speculate it is due to natural variation in the temperature of circulating deep ocean currents. And, indeed, there may well be other, longer-term cycles.

American meteorologist Warren White and his Australian colleagues have shown that the global climate system probably consists of a series of interacting 'waves' that create interference patterns at many timescales. The Inter-decadal Pacific Oscillation may just be an expression of one of these patterns. These patterns involve cycles of 20 years or more that have an impact on many desert regions, including those of Australia. One day we may be able to forecast these with more confidence, but for the moment our records are too short to work out their complex patterns. The long time frames of

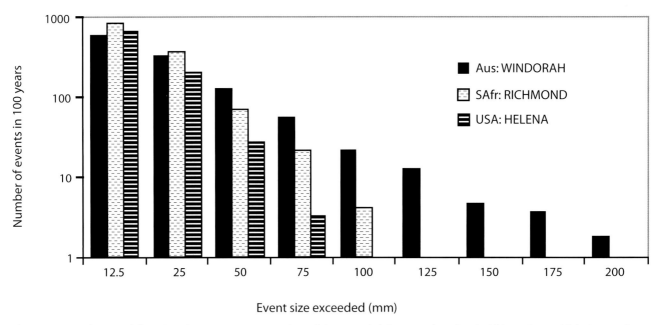

Figure 1: Australian rainfall stations have many more, and much larger, rainfall events than South African sites, which, in turn, have more than those in the USA. The histogram compares the size of rainfall events over 100 years at three weather stations with roughly the same seasonality and average annual rainfall (all about 300 mm y^{-1} with two-thirds in summer) – Windorah in western Queensland in Australia, Richmond in South Africa and Helena in the USA. There are many events of at least 12.5 and 25 mm at all stations, but only Windorah has events of larger than 125 mm, and its extremes are sometimes larger than 200 mm.[34]

these desert cycles mean that an individual human can never experience the full range of possibilities, and hence it is hard for desert dwellers to learn about them in one lifetime.

Besides low and unpredictable rainfall, Australia's deserts also exhibit extremes of rain, floods and droughts. Major flooding rains are important because, after filling the soil, they have water left over to recharge watertables and to fill wetlands and salt lakes. If unchecked, that water can also sweep across and erode the landscape.

Big rainfall events may consist of very wet days, weeks or even years that affect large areas. Rainfall distribution graphs of Australian weather stations (Figure 1) show a 'long tail', which represents occasional huge rainfall events that do not occur in most other regions of the world. When they occur over large areas, these big events recharge groundwater for years to come, and allow the seedlings of big shrubs and trees to get their roots down deep into the groundwater before the surface soil dries up.

In addition to these large regional events, there can be extreme local downpours that produce major floods. These local falls are a consequence of the high energy of desert climates, and thus occur in most desert regions. They are caused by powerful storm cells delivering intense rain on to an already saturated catchment that has little vegetation to stop run-off. When perfectly positioned, they cause what CSIRO geomorphologist Geoff Pickup calls 'mega-floods'. In the Finke River in central Australia, he found old sediments indicating floods far greater than anything that has occurred since the first Europeans arrived in the 1850s. He dates these giant floods to about 5000 and 8000 years ago – providing a hint about how often these mighty events occur,[35] as well as a warning for towns and communities that may be in their path.

For example, on 18 January 2007, Alice Springs hydrologist John Childs noted a big storm 200 kilometres from the town, which dropped 246 millimetres of rain on Numery Station.[36] This was equivalent to its normal annual rainfall falling in one day, with

215 millimetres falling in less than 6 hours. Childs observed that the Alice Springs flood in 1910 is usually described as the largest since European records began, with an 80 year return time. A major flood at Easter in 1988 had only about a 40 year return time; even so, it overflowed the river banks and sent about 10 centimetres of water down Todd Mall in the town centre. It killed three people in the riverbed and caused millions of dollars of property damage. By comparison, had the 2007 storm sat over the town's catchment instead of over Numery Station, 'the likely result from hydrological models would have been a flood with a 100 to 200 year return time', says Childs. This exceeds the maximum used for town planning, and the Mall might have been under two metres of water. In such an event, the town would be devastated, and numerous deaths could be expected.

Deserts also experience extreme drought periods and their inhabitants must be hardened to this. When desert plants grow actively after rain, many make a trade-off between growing fast in the short term then seeding, or investing in roots to last through the coming dry time; the trade-off that works best depends on when the next rain is likely to fall. To plants, their water supply is a series of wetting events and drying periods. Some events are big and some are small; some of the dry periods are short and some are long. What really affects desert plants are the return times for events of different sizes, which define the lengths of dry periods that they must endure. The return time for a 50 millimetre rainfall event, for example, is the average number of days after one of these until another occurs. The patterns of these return times are also unusually variable in Australia, due to the interaction of the different climatic cycles noted above. As for its rainfall distributions (Figure 1), Australia has a bigger tail in the distribution of time between rains than most other regions[37], which means that every now and again very long dry periods occur. These can give rise to droughts the like of which non-Aboriginal Australians have absolutely no experience in their brief 230 year tenure.

Responding to such uncertain droughts requires very different strategies to dealing with reliable droughts, however harsh. For example, cacti thrive in the central American deserts, where part of the year is very dry but the wet season is quite reliable. Cacti are succulents: plants that store water in their huge stems to keep them going during a dry period. With confidence that rain will come every year, and the store can be recharged, this strategy works well. But plant ecologists in Australia such as Noel Beadle in the 1940s soon realised that there were very few succulent plant species in Australia (Plate 5a), and it is easy to see why. Sooner or later a succulent in Australia will be faced with a dry period that is too long for its internal water store and it will die out. Australian plants cope with such variability using other strategies, which we will examine in Chapter 3.

We are now entering an era when large tracts of the planet are expected to become drier, more extreme and more unpredictable. In this future world, climate events that have not been recorded in human history will become ever more common. Like desert plants, we will need risk management strategies that cope with these uncertainties, rather than relying on old rules of thumb.

We are, in a sense, all becoming desert explorers.

2.3 Climate trends ahead

The world is presently in an interglacial period; were it not for global warming, this would eventually end in another ice age. In the past, interglacials have been both warmer and wetter than the glacial periods – favouring the growth of forests and the contraction of deserts and grasslands. Indeed, before the advent of humans with axes and fire, most continents were extensively forested. In a world without humans, it is highly probable that the forests overall would now be continuing to spread at the expense of grasslands, and open woodlands and grasslands would be encroaching on deserts. Within this broad trend, some parts of the world – mainly in the mid-latitudes such as Australia – would still have tended to dry out during the last 6500 years, due to the increasing frequency of El Niños and to lower monsoonal rainfall.

Today, though, few would disagree that humans are changing the trajectory of climate. Coupled with human activity directly affecting the drylands, this introduces a wild card into the naturally uncertain process. Indeed, the effects of humans on our planet are reversing the natural trend of interglacial periods on a broad front, by reducing the area of the world's forests and grasslands and expanding the deserts and drylands.

In its 2007 assessment, the Intergovernmental Panel on Climate Change warned that global warming is now beyond dispute and that it is 'very likely' that most of the warming observed in the last 100 years is attributable to human activity through the release of greenhouse emissions. The warming trend of the past 50 years is 'very unlikely' to have been a freak natural occurrence. Their report estimated that the Earth will be 1–6°C warmer, on average, by 2100, depending on how humanity responds to the challenge – at the lower end if we reduce emissions quickly, at the higher end if not. The global mean temperature is already up by about 0.74°C compared with the early 1900s, with a strong trend since around 1970. The warming will be most marked over land, and in the high latitudes, and will be accompanied by more noticeable changes in the occurrence of extreme weather events, including protracted droughts. This process is now likely to persist for centuries, even if immediate and widespread action is taken to rein in carbon emissions.

The impact of climate change on the world's deserts is, as yet, far from clear. Although scientists agree that global warming will bring more rain on average worldwide, many deserts are likely to become hotter, drier and more prone to extreme events, including floods, fires and droughts. The evidence for this lies in weather records from 1900 to 2005, which already show that the African Sahel, the Mediterranean, southern Africa, and parts of southern Asia and Australia have all become drier; this trend has accelerated in the last 50 years (see Figure 2).

Since the 1970s, warmer conditions, higher evaporation rates and decreased rainfall in many deserts have contributed to more intense and longer droughts. There has also been an observed increase in the intensity of tropical cyclones – one of the main sources of rainfall for Australia and its deserts – although what this portends in terms of actual volume of rainfall is not clear. A further complication is that the warming temperatures drive more evaporation from a given amount of rainfall.[38] Even though rising CO_2 enables plants to use water a little more efficiently, the net effect is likely to be harsher growing conditions.

The United Nations Framework Convention on Climate Change states that 'countries with arid and semi-arid areas or areas liable to floods, drought and desertification … are particularly vulnerable to the adverse effects of climate change'. The secretariat of the UN Convention to Combat Desertification added in 2008 that 'Scientists cannot yet predict how rising atmospheric levels of greenhouse gases will affect the global rate of desertification. What they can predict is that changes in temperature, evaporation, and rainfall will vary from region to region. As a result, desertification is likely to be aggravated in some critical areas but eased in other places'.[39] In short, the uncertainties that characterise deserts are set to increase.

Recent climate modelling by the British Hadley Centre points to the disturbing possibility of a major increase in the frequency, intensity and extent of drought through the twenty-first century. Running their climate model against actual weather events from 1952 to 1998, Hadley researchers found it simulated the frequency of past droughts well. In line with reality, the model showed that 'the percentage of the land surface in drought had increased by the beginning of the twenty-first century from 1% to 3% for the extreme droughts, from 5% to 10% for the severe droughts and from 20% to 28% for the moderate droughts.' This supports the view that the climate has been drying, as well as warming, over the last 50 years.

But when the Hadley Centre researchers started to look forward in time, they went on to say: 'This increase continues throughout the twenty-first century and by the 2090s the percentage of the land area in drought increases to 30%, 40% and 50% for extreme, severe and moderate drought

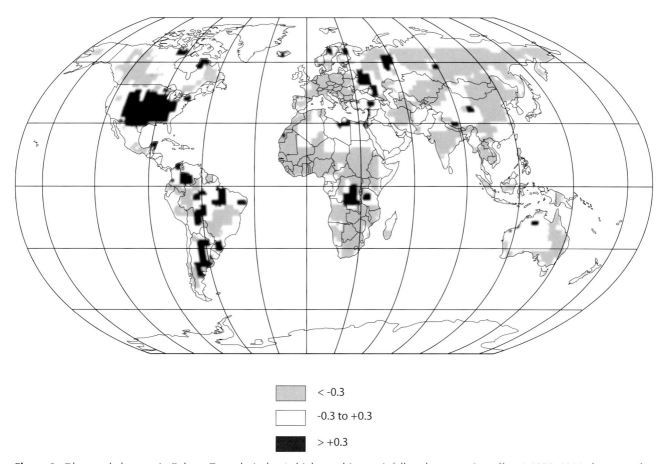

Figure 2: Observed changes in Palmer Drought Index (which combines rainfall and evaporation effects) 1952–1998 shown as the change per decade: grey areas have experienced increasing drought, black areas increasing moisture – the greys are clearly more widespread than the blacks.[40]

respectively.[41] In other words, moderate drought conditions that used to afflict 20 per cent of the world in any given year may be affecting half the world by 2100.

What does all this mean for Australia in particular?

As suggested by the global observations, more detailed studies by Australia's Bureau of Meteorology show that the northern and western half of the continent has become wetter since 1950, while the southern and eastern half has become drier (see Figure 3). Droughts have become hotter and therefore more intense. From 1950 to 2005, extreme daily rainfall has increased in north-western and central Australia and over the New South Wales western tablelands, but has decreased in the south-east, south-west and central east-coast. According to recent work by CSIRO and the

Bureau of Meteorology, very extreme hot periods are becoming more common in most regions, and there are more years of very low rainfall in some regions, particularly south-western areas.[42]

The Intergovernmental Panel on Climate Change's broad outlook for Australia (Table 1) under future climate change is for decreases in annual rainfall of 5–50 per cent, accompanied by 1–8°C of warmer temperatures.[43] Within that broad trend, however, some areas, such as the north-west, may experience increased rainfall, while others, such as the centre and the farming areas of the south, east and south-west may find the climate becoming drier and more erratic. It is probable that the inland pastoral zone will expand into the surrounding wheat/sheep belt, that winter grain cropping will retreat on to areas with better soils and into the more reliable rainfall near the coast, and that the summer

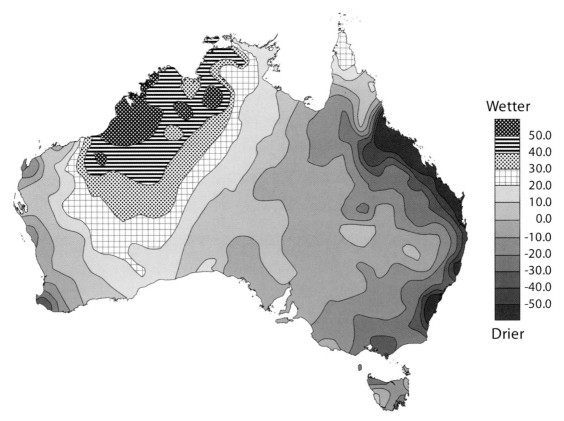

Figure 3: Measured trend in rainfall for Australia from high-quality Bureau of Meteorology weather stations, 1950–2008, showing wetting (positive trend) in north-western Australia but drying (negative trend) in the south-west and eastern parts during this period.[44]

rainfall belt will forge its way south at an average rate of about 10 kilometres per year. All areas will be prone to more extreme events – droughts, fires and floods. There are concerns that sea surface temperature patterns characteristic of El Niños may occur more often and last longer. Climate models suggest that up to 20 per cent more droughts[45] will occur over most of Australia by 2030 and as much as 80 per cent more droughts by 2070 in south-western Australia.

On the face of it, these predictions point to accelerated drying for the Australian continent as a whole and to expansion of the deserts and drylands in the area, although this may be offset by higher rainfall in some regions. This sort of change sounds like it might be impossible to handle. But, of course, some regions of the world have always had to deal with it – the deserts!

2.4 State of the world's deserts

In 2000, the United Nation's Millennium Ecosystem Assessment[46] – a stocktake of the world's environments commissioned by then Secretary-General Kofi Annan – established that drylands presently occupy 41 per cent of the Earth's land surface, and the desert component about 33 per cent. The latter are home to about a fifth of the world's population, as well as countless valuable and unusual animals and plants (see Box 2). About half of these people depend directly on natural resources for their livelihoods and day-to-day survival. They include most of the world's rural poor, whose birth rates continue to be high relative to advanced countries and to cities. If present population trends are sustained, it is likely that natural increase alone will lift desert populations close to three billion by mid-century. At the same time, the physical

Table 1: Latest predicted trends in climate for different areas of Australia.[a]

Region	2020	2050	2080
Temperature change (°C)			
0–400 km inland of coast	+0.1 to 1.0	+0.3 to 2.7	+0.4 to 5.4
400–800 km inland	+0.2 to 1.3	+0.5 to 3.4	+0.8 to 6.7
Central Australia	+0.2 to 1.5	+0.5 to 4.0	+0.8 to 8.0
Rainfall change (per cent)			
Within 400 km of western and southern coasts	−15 to 0	−40 to 0	−80 to 0
Sub-tropics (latitudes 20–28°S) except west coast and inland Queensland	−10 to +5	−27 to +13	−54 to +27
Northern New South Wales (NSW), Tasmania and central Northern Territory (NT)	−5 to +10	−13 to +27	−27 to +54
Central South Australia, southern NSW and north of latitude 20°S, except central NT	−5 to +5	−13 to +13	−27 to +27
Inland Queensland	−10 to +10	−27 to +27	−54 to +54

[a] Table 11.4 in Hennessy et al. (2007); readers can readily explore projections for themselves (for slightly different future dates) at <http://www.climatechangeinaustralia.gov.au/index.php> (accessed March 2009).

area of the deserts themselves is expanding. It is on the cards that their human population will exceed four billion by the second half of the twenty-first century simply by virtue of the deserts' burgeoning geography.

One of the most significant features of deserts worldwide is that most of their inhabitants are poor. Ninety per cent of their people live in developing countries and experience far lower levels of health care, education, communication and service delivery than the average citizen of the planet. They are thus among the world's most vulnerable to disease, hunger and infant mortality. Any tendency of deserts to expand or dry out further will exacerbate these problems.

Nevertheless, deserts (and drylands more generally) are a vitally important component of the world economy, providing benefits far beyond their own inhabitants.[47] They are home to 80 per cent of the world's 3.3 billion cattle, sheep and goats and yield most of its mineral wealth. They account for 43 per cent of the world's cultivated farmland in the less arid parts and thus furnish a very significant part of its grain supply. They produce many valuable commodities that create trade and employment in cities and temperate regions. Their soils contain about a quarter of the world's soil carbon and almost all of the inorganic carbon. Deserts are a primary source of the nutrients that sustain humanity and

– contrary to what might seem obvious – they are huge exporters of water, either directly in rivers and irrigation channels, or else embodied in the meat, timber, grains, metals, tourism, energy and other products and services they provide to people outside desert regions. For example, it takes as much as 50–100 000 litres of embodied water to grow the pasture or feed that sustains an animal for each kilogram of meat that it produces and 120–170 000 litres of water to produce a kilogram of clean wool.[48]

Because it evaporates so freely in warm climates, water is scarce in deserts, and is becoming scarcer with misuse and mismanagement. In the past half-century, global water use has grown at a rate of 25 per cent every decade, and current demand from the world's cities is expected to double within a generation. The effect of this soaring demand is to reduce still further the amount of water available to people, plants and animals in deserts. The Millennium Ecosystem Assessment predicted that water availability per person in these regions will decline to 1300 cubic metres a year, which is only two-thirds of what is regarded as the minimum for human wellbeing. This growing water stress will, in turn, limit the local production of meat, grains and vegetables: increasing the risks of poverty and hunger. For example, the International Water Management Institute has estimated that declining groundwater levels in India could reduce the grain harvest by as much as 25 per cent at a time when both population and consumer demand are expected to soar.[49]

The UN considers that there is a strong likelihood that changes in the global climate and vegetation cover – whether natural, the result of human activity or both – will bring about a further decline in water availability and biological production in these dry regions.

Indeed, the combination of human activities such as overgrazing, over-clearing of land, over-extraction of ground- and surface waters leads to desertification – the enlargement of the deserts – which could intensify with climate change. Accurate estimates of global rates of desertification are hard to come by, because different measurement methods are used in different parts of the world and, in any case, distinguishing permanent change from the natural variability of deserts is difficult. However, the UN Environment Program (UNEP) considers that desertification now affects up to one-third of the Earth's land surface and more than one billion people.[50] The land area affected by desertification world-wide is estimated at between 6 and 12 million square kilometres: one to two times the total area of Australia.

'Nearly one-third of the world's cropland has been abandoned in the past 40 years because erosion has made it unproductive. Each year an additional 20 million hectares of agricultural land either becomes too degraded for crop production, or becomes lost to urban sprawl', UNEP says.

Desertification affects about a third of irrigated farmlands, almost half of rain-fed lands and three-quarters of all grazed rangelands. Annually, about 2 million hectares of irrigated land, 4 million hectares of rain-fed cropland and 35 million hectares of rangelands are estimated to lose all or part of their productivity. The consequences of this process extend far beyond food and water production to affect livelihoods, health, education, birth rates, refugees and armed conflicts.

The Millennium Ecosystem Assessment also notes that it is far easier to prevent desertification than it is to reverse it, because it can take a thousand years to replace the 5 centimetres of soil that may be lost in a single dust storm. 'Population pressure and bad land management practices are the cause of degradation. Better management of crops, more careful irrigation, and strategies to provide non-farming jobs for people living in drylands could help to address the problem', it says.[51]

UNEP warns that failure to tackle this complex of problems may lead to people in their millions leaving their homes and fleeing to other parts of their region and worldwide in search of food, shelter and security. Our view is that one of the best investments Australia can make in its own security is to help our neighbours to avoid these dryland pressures. Finding positive futures for our own desert regions is a key strategy towards doing this.

2.5 Responding to uncertainty

The scenarios of rising future drought and uncertainty are very disturbing. Not only do they imply a significant increase in the extent of deserts, but also further desertification of those regions that are already dry. This has real consequences for the billions of future inhabitants of the world's deserts, as well as for total global food supplies. Regions that are currently net exporters of food to the world's great cities and temperate areas may well be unable to supply them, causing shortages even in the most advanced and developed countries. And regions which are currently agriculturally marginal may become the scenes of food crises. Without preparation, this could displace streams of refugees into the cities and to the developed world on a scale that dwarfs the greatest migrations of history.

As the vegetation cover in the desert shrinks from the combined hammer blows of local overuse and climate change, more carbon will be released into the atmosphere. The capacity of the Earth's drylands to act as a carbon sink will be proportionately lowered, thus offsetting efforts to soak up carbon by tree planting or other means elsewhere. Unless it can be checked, the expansion of the world's deserts will thus accelerate and magnify the impact of global climate change.

Such a scenario would be disturbing indeed if humanity sat around and did nothing about it. However, because they are difficult places in which to make a living, deserts often bring out the best in humans through many technological and social advances. At least, they have done so often in the past and there is no reason that they should not again.

The challenges of desertification and climate change may, in fact, yield a fresh outpouring of human imagination, creativity and innovation. Australians, being masters of drought and dry landscapes, are well placed to lead in this new chapter of the history of humans on this planet – a chapter in which desert knowledge has the potential to play a significant role.

3

What drives deserts?

'In short, one should treat deserts as deserts and not try to apply non-arid standards to their development except under very specific circumstances.'

AMIRAN 1973[52]

Deserts are different. Although they may remind us of familiar things and places, in fact they operate in ways that are far different from what we might expect. The creatures and plants that inhabit them are engaged in a life-and-death struggle for perilously scarce resources that shift both in space and time, and are therefore masters in survival.

This means they have much to teach us about our livelihoods and our settlements. Our environments – deserts or not – are coming under increasing stress and increasingly fluctuate in unpredictable ways as a result of the ceaseless pressure of human population and its demand for resources. Yet this type of uncertainty is at the base of how deserts operate.

3.1 The desert drivers

The three-quarters of Australia that we loosely term desert possess some essential features shared with desert regions elsewhere in the world, but which differ from non-desert areas. These features drive how deserts function, both physically and socially – they are 'desert drivers'. They are also causally linked, and understanding these links is central to understanding how deserts work – and

how many other places will work as they become more desert-like.

The variability, unpredictability and extremes of the desert climate result in low and very variable primary productivity. This means that much desert life also fluctuates wildly, in a series of booms and busts. In the case of Australia, this variation is exacerbated by ancient impoverished soils on a flat, worn-down continent. Water and wind sweep a thin dust of nutrients across the landscape. Where this 'gold' lingers for a time, it enriches the local area, causing plants and animals to flourish. Then another flood or dust-storm sweeps it elsewhere, and life dies down to a narrow subsistence, or even vanishes.

All this makes the productivity of deserts extremely patchy, but it also makes them diverse because many organisms have adapted to use this patchiness to their advantage. The low productivity means that human populations, too, are generally sparse, patchy and often highly mobile. The deserts are thus the natural realm of the nomad, the hunter-gatherer, the drover and the prospector who obey the same rules of survival.

In the modern age, this sparseness results in populations too low to sustain a critical mass:

markets for goods and services are remote, as are centres of political, economic and social power. As a result, desert peoples often find themselves without a voice in the larger society. These factors, exacerbated by turn-over in the small pool of labour, create an extra layer of unpredictability in markets, policies and labour, which is generally outside the control of desert dwellers. They also mean there is comparatively little research into the needs of these regions and their people – although plenty of local innovation occurs as a matter of necessity, as we shall see in later chapters. The scale of management required for landscapes, business networks and connections to regions outside the desert means that local knowledge assumes particular significance, although it is also hard to develop in such a variable environment.

Remoteness conveys advantages too: because of the sparse population, traditional culture has persisted in desert regions while it has declined elsewhere, and Aboriginal local knowledge is often particularly significant. Furthermore, the type of people who like to live in these conditions tend to exhibit particular cultural and social characteristics, either in the long term (because they were born into an Aboriginal desert culture, for example) or because they are the ones who visit and choose to stay.

The links between the desert drivers – unpredictability, scarce and patchy resources, sparse populations, mobility, remote markets and isolation from political power, cultural differences, local knowledge and social uncertainty – are illustrated in Figure 4. They are what characterise deserts, and make them different (though they apply more or less to some other environments too – see Box 4). These drivers are very important to understanding the message of this book, so we will explore them all further in this chapter.

3.2 Unpredictable climate and scarce resources

Chapter 2 looked at how variable, extreme and unpredictable the climate of desert Australia is. Low and unpredictable rainfall is a fundamental contributor to the uncertainty in production, but

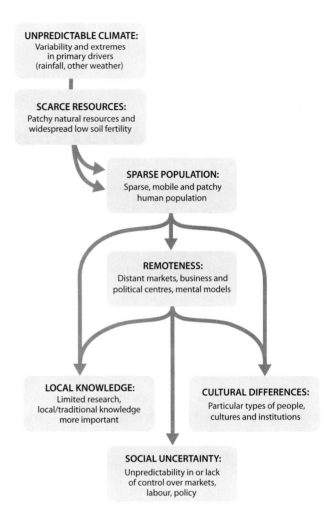

Figure 4: The 'desert drivers' that critically determine how deserts work.[53] These are causally linked: the unpredictable climate and scarce resources result in a relatively low population. As a result, people are remote from markets, centres of business and power, and the places where most mental models of how things work are developed. Given limited numbers of people in the deserts and their remoteness, local knowledge is particularly important and certain types of culture tend to predominate. All these factors combine to create further 'social uncertainty' in the socioeconomic realm, such as in prices set on distant markets, policy set in distant centres, and the turnover of people in small communities having a disproportionate effect on the skills base.

this is exacerbated by the fact that desert soils are generally low in nutrients. In turn, this is partly because there is little water to grow plants so there are fewer leaves shed with which to build up soil fertility. However, the great deserts are also in the middle of ancient continents – the oldest and most worn-down rocks of Africa, Asia, North America

BOX 4: HOW WIDELY DO THE DESERT DRIVERS APPLY?

The desert drivers of Figure 4 operate acutely in the remote, dry and variable deserts of Australia, which are the focus of this book. However, there are obviously other parts of the world that are remote or sparsely populated and there are some much more densely populated desert regions. How general are the ideas?[54]

Unpredictable climate: this is true for many deserts, but exacerbated where El Niño affects the climate; however, extremes of cold in winter (e.g. Patagonia, Kazakhstan and northern Canada) can be just as variable, and extreme events such as cyclones and bushfires can have high impacts (but are less continuous).

Scarce resources: true in almost all deserts, but also in other areas of extremes, particularly near the Arctic and Antarctic.

Sparse population: relatively true in most deserts and other resource-limited regions, though less true in subsistence societies than developed countries; the impact may be locally reduced by exploiting major rivers for water (e.g. the Colorado in the USA) or mining for financial subsidies (e.g. oil in the Gulf States).

Remoteness: true in all areas with sparse population, though less significant if a whole country is sparse (e.g. Mongolia and Libya) than if part of the country is heavily populated (e.g. Australia and many sub-Saharan countries). Also true for inaccessible rainforest regions (e.g. Brazil or Papua New Guinea, though these are being developed rapidly) and for remote islands (as in the Pacific, where remoteness is determined by ocean isolation).

Local knowledge: particularly important issue where the population is remote from decision-making centres; hence, true in most remote deserts but also in remote rainforests and islands where traditional cultures have also often persisted.

Cultural differences: most likely to matter where the population is remote from decision-making centres, as for local knowledge.

Social uncertainty: common in deserts, but driven primarily by the effects of remoteness and hence occurs in other remote areas.

Clearly a sparse population can have various causes, but these usually involve some sort of resource limitations; however, given a sparse population, remoteness almost always ensues, and significant social uncertainty is usually a consequence of this. In some cases, it is *relative* remoteness (and *relatively* sparse populations) compared with surrounding regions of greater political and economic power that matters. Many of the messages of this book are relevant wherever there is uncertainty in climate and social systems, and relative remoteness from population centres.

and Australia. In regions such as the Aïr Massif in Niger, the hills are the broken down remnants of scree slopes that were themselves formed by the erosion of more ancient mountain ranges. The elemental nutrients trapped in the original rock have been scrubbed away, again and again, over many millions of years. In these areas, the shallow soils and the sands that blow across the landscape are so poor to start with they can do little to generate plant growth. They are exhausted. In a recent paper, researchers Gordon Orians and Tony Milewski argue that intense fire has also contributed to the loss of key micro-nutrients such as zinc, iodine, cobalt and selenium from Australian soils, further exacerbating the poverty of production for animals that need these elements.[55]

In Australia, about two-thirds of the desert area has soils like this – phosphorus-poor red earths and blowing sands (Box 1). The few nutrients these soils contain are mostly squeezed into the top one or two centimetres, and it takes little for them to be blown or washed away. However, if water washes over them and then stops a little down-slope, the nutrients may concentrate in a small patch (Figure 5). Geomorphologist Geoff Pickup has coined the term 'erosion cell' for these small, repeated patterns on intact flat landscapes where water flow has picked up a little soil, with its nutrients and seeds, and dropped it again a few tens of metres away in a depositional area, or 'sink'.[56] This pattern can be repeated all down a shallow slope, creating many patches with low and high fertility. As he says, 'once sediment is deposited in a sink, it provides a favourable environment for plant growth because of its moisture status, nutrient content, seed supply and relative stability compared with other parts of the erosion cell'. The plants drop leaf mulch that, in turn, helps to trap more water and nutrients. This process 'self-organises' into the patchy landscape so characteristic of the desert – places of intense and wildly varying fertility amid generally depauperate surrounds. This is the key to the richness and diversity of desert life. If you spread the same amount of water and nutrients evenly across a desert landscape, it would not support even a fraction of the plant and animal life that this patterning does.

Patchiness is one of the great secrets of Australia's biological success.

Not all desert soils are poor. About a third of desert Australia sits on richer soils. These occur in places once inundated by a sea that left behind calcareous clays, or by large lakes during wetter times when fine black clays were formed, both of which retain moisture and nutrients. These regions are flat and resistant to erosion. Instead of erosion cells,

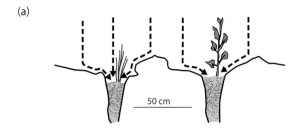

(a)

50 cm

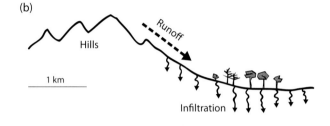

(b)

Hills

Runoff

1 km

Infiltration

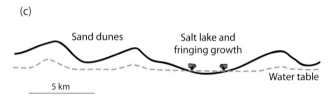

(c)

Sand dunes

Salt lake and fringing growth

Water table

5 km

Figure 5: Scarce desert water resources get concentrated in patches at various scales, often taking scarce nutrients and seeds with them. The scales include (a) cracks among rocks; (b) the foot slopes of small hills; and (c) large-scale watertables and drainage lines in sand dune deserts.[57]

their 'cracking clays' tend to form scattered depressions called gilgais that appear when wet soil swells, then cracks and shrinks as it dries out (Plate 4d). These depressions accumulate extra water and nutrients and become richer patches. Other types of clay soils in southern Australia carry saltbush and bluebush shrubs. Here, it is the plants themselves that act as traps for windblown nutrients and seeds, creating patches of fertility in a poor environment.

Other factors also contribute to the many patterns of deserts. Wind-blown dunes create their own biological forms, because the deeper sand holds water long after moisture has vanished from the surface. Some plants have adapted to exploit this. Elsewhere, unpredictable storm cells leave wet patches that become temporary magnets for life from the surrounding dry country. Hail storms can strip leaves from swathes of woodland and mulch

BOX 5: SELF-ORGANISING DESERT PATTERNS

If you accidentally drop a handful of rice on the kitchen floor, it is very slow and inefficient to pick it up again grain by grain. It is far easier and quicker to sweep the grains together then pick them all up in a dustpan. This is because the grains are like a scarce scattered resource: there is only enough of them to collect usefully where they have been concentrated together. The same is true for water and nutrients in deserts: if they are scattered randomly across the landscape, there is never enough in one place for a tree or a large herbivore to use. Such organisms may therefore only live in refuge areas created by topography, but plants themselves can also become organised to concentrate these resources.

The classic example is the mulga grove (see diagram below and Plate 4c), where chance clumps of trees start to capture water flows across the landscape.[58] The plants growing in the areas where resources are concentrated grow better and produce more litter, which helps to capture more of the next flow. Meanwhile, the intervening areas remain bare and lose water instead; this water supplies another clump further downslope. So, the pattern becomes self-reinforcing. It may take centuries for the pattern to get established, but, once it is, it is quite stable. It is what scientists call a 'self-organising system' – it does not require any external help to happen.

Repeating pattern of mulga groves

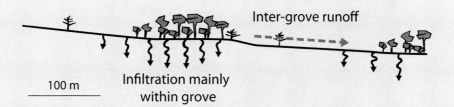

The patterns form at all sorts of scales in resource-limited environments – bands of grass across slopes like miniature mulga groves; the 'tiger bush' (brousse tigre) of west Africa, which is exactly like the mulga; bluebush shrubs on flat landscapes where they intercept wind-blown material rather than water; as well as large-scale erosion cells, where water courses flood out in a patch of dense vegetation, then erode out again on their downslope side, with the pattern repeating across the landscape.

the desert. Fire paints the landscape with patches of re-growth of varying age. All the drivers conspire to make the desert patchy.

In short, where resources are poor, all sorts of processes create local concentrations that support life that could not exist under the average conditions across the whole landscape (see Box 5).[59] Sometimes these patches are huge – hundreds of square kilometres in extent. Those caused by water movement tend to be smaller: from a few square metres to a few hectares. Some are tiny islands of richness, formed round the stem of an individual shrub or tree and extending no more than a metre or so into the barrenness around it.

All these patterns have immense significance to life in deserts, and to the humans who live there. Though the reasons are different, our use of deserts for mining, pastoralism, tourism and community life reflects a similar patchiness and scatter of resources. This is pivotal to the way Australians must think about the future in a generally arid continent in increasingly uncertain times.

3.3 Sparse settlement patterns

Deserts usually support far smaller human populations than neighbouring regions that are less arid. This seems obvious, due to the low productivity of the environment and lack of water. The world's most populous regions mainly began as agricultural areas in fertile river basins that gradually became centres of population and cities that, in turn, attracted more and more people. By comparison, desert regions started with low populations and have mostly stayed that way.

The few exceptions to this rule are places such as the Arabian Gulf States, which have converted the resource of oil into water for development, and Phoenix, Arizona and Las Vegas, Nevada, which depend on water and investment from outside the desert. These cities each depend on a particular economic option that is not widespread in the world, and may not be sustainable in the long term.

The population in most deserts is concentrated in major service centres surrounded by smaller specialist towns, which are themselves surrounded by small settlements close to the patches of productive resources on which they depend (Figure 6). In Australia, the overall arid zone population density is less than 1 person per 20 square kilometres[60], which

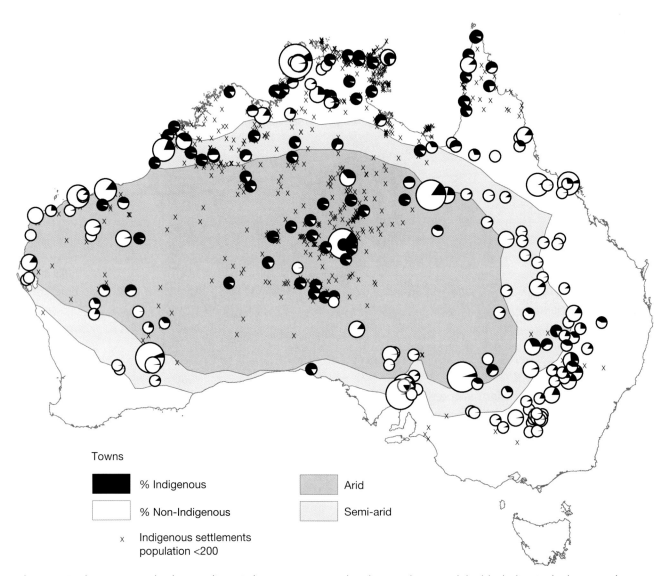

Figure 6: Settlements in outback Australia – circles are proportional to the population and the black slice in the larger settlements shows the proportion of Aboriginal inhabitants.[61] Note how the larger towns tend to cluster around the periphery of the desert. True desert settlements in Australia really only consist of a few larger, mixed enterprise service centres, and then a large number of more or less single purpose small settlements in their hinterlands. This creates the need for approaches to remote governance and services that are more or less unknown outside deserts today.

is comparable to the remote Yukon and Northern Territories of Canada. By comparison, the average square kilometre in Australia contains about 2.8 people, in central Sydney it contains over 4000 people, and cities in India are approaching 10 000 people in the same area!

The major service centres, such as Alice Springs, Kalgoorlie, Broken Hill and Mount Isa, typically contain 15 000–30 000 people; smaller towns, such as Longreach, Yulara, Newman, Coober Pedy and Bourke, contain a few thousand people, and are surrounded by what would be called villages in other countries of less than a thousand down to only twenty or so people.[62] The larger towns tend to be dominated by non-Aboriginal people, while smaller settlements tend to be either more than 75 per cent Aboriginal or more than 75 per cent non-Aboriginal, with very few truly mixed (Figure 6).

Most of the world's deserts have higher average population densities than those of Australia, but have a similar structure of larger service centres, intermediate regional centres and small outlying 'villages'.

Deserts favour mobility and Australia's desert population is highly mobile – in three different ways. Firstly, there are the 'transient residents' – this includes 'fly-in-fly-out' miners, consultants, politicians, business representatives and tourists; in the 2001 census, one in seven persons recorded in the arid zone did not regard it as their normal place of residence. Secondly, there is a substantial (mostly non-Aboriginal) population of short-term residents – teachers, medical staff, jackaroos and public servants – coming for perhaps 2 years 'hardship posting' before resuming their family ties and careers in capital cities. Of course, some of these fall in love with desert life and never leave. However, a quarter of the non-Aboriginal population of the desert in 1996 had turned over by the 2001 census (about 116 000 people) – actually leaving the desert – compared with only a tenth of the Aboriginal component (about 8500 people) doing the same.

The Aboriginal population is the third, and a most significant, category. They are very mobile, but, unlike the previous two groups, their movements are mainly within the desert region.[63] Their

activity is more like the classic nomadic movements of other deserts and bygone days, albeit now carried out using modern transport and communications technology and infrastructure. A small number of non-Aboriginal people also pursue this type of mobility – miners moving between mines, some teachers, doctors and storekeepers who are committed to the desert, shearing teams, shooters and dog fence minders, ecotourism operators, and even the odd scientist.

Thus the population is as sparse and patchy as the underlying resources, and it is mobile to deal with this. Imagine planning a city where a quarter of the population has moved on in 5 years. Overall, though, the sparseness means that people are remote from one another and from the major centres, which leads to further implications.

3.4 Distant markets, distant voices

The world's great marketplaces are in the megacities, far from the deserts. There, too, government and large corporations place their headquarters. If you live in a desert, you rarely encounter any of them. They are remote, lordly and seldom pay close attention to the interests and needs of desert dwellers. In Australia, the rural–urban divide is striking and the outback–urban divide even more so. Although the arid and semi-arid zone is home to more than half a million people (nearly twice the population of the Australian Capital Territory, which contains the national capital, Canberra) there is no seat of government in the desert, no main university campus, no corporate headquarters, no stock exchange, no major media organisation, and not even a central livestock saleyard. This reflects the fact that population density, rather than total population, is what really counts for creating centres of influence and demand. Here desert dwellers are at a disadvantage.

Graeme Hugo and colleagues at Adelaide University developed the Accessibility/Remoteness Index of Australia to express this remoteness – and the disadvantage it imposes on people in their access to services.[64] Their remote categories are shown on Figure 7. Essentially, all desert Australia lies in their remotest category, other than a few small pockets

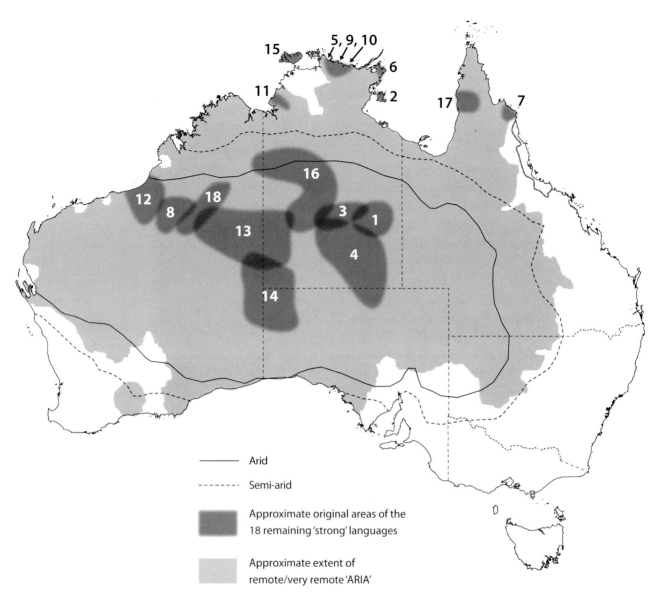

Figure 7: The extent of the remote categories of the Accessibility/Remoteness Index of Australia (ARIA) superimposed on the map of desert Australia (cf. Box 1). The approximate regions of the 18 remaining strong Aboriginal languages (numbered) are superimposed, from which it is apparent that, although not all remote regions have retained strong language, all strong languages are to be found in remote regions.[65]

around the big service centres. The real challenge for desert Australia is to come up with ways to turn this apparent disadvantage into an opportunity. To do so means getting around the vast gravitational pull of economic centralisation.

Historically, when transport and communication were poor, small towns had to sustain themselves. People could not seek food or services elsewhere. Isolated towns such as Alice Springs and the small

gold rush settlement at Arltunga had far more facilities than villages in more closely settled parts of the country. As transport links improved, it was possible to concentrate production where it can be done more cheaply – so market gardens, banks and other services disappeared from small towns. Some things have to be local, though – people need haircuts and water supplies locally and regularly, and you cannot visit Uluṟu anywhere but at Uluṟu.

At some point, however, improving communications technology means that people can buy and sell just as well from a remote location as they can in a city, so some forms of business can prosper in remote regions where geography is no longer a barrier. But the kind of business is important: low-value items that need a large market will seldom find it in a desert. If you have to truck a bulky product such as bricks 2000 kilometres to a city market, then you'll never compete with the company that makes the same items in that city. But if you have a high-value product that is in demand and costs little to ship (such as knowledge, art or even gold), and that preferably can be honed in a small local market before reaching out to cities, then the options change dramatically.

Thus the competitive advantages of living in the desert in this fast changing world are to be found around: the resources that only occur in the desert (particularly natural and cultural resources, such as mining, nature tourism and art); the innovative products that deal with the challenges of living in the desert but which may be viable elsewhere once those challenges have been overcome; or providing local services that cannot be delivered from elsewhere.

A more subtle impact of distance is on standards and services. In many regards, desert dwellers are required to meet the same standards as densely populated areas, simply because the standards are developed where most people live. These often turn out to be quite inappropriate to desert conditions. Few governments or large companies think it worthwhile to tailor their services to local conditions and needs in deserts. Thus, Alice Springs has basically similar items on supermarket shelves as in Adelaide, leading to occasional oddities like bins of cut-price rain-coats when it is wet down south but 40°C in the Centre. Parallels may be found in education and training, health and social security services, where the same approach is applied across the country. This one-size-fits-all approach creates significant problems. For example, the failure of health care to resolve the crisis in Aboriginal health is at least partly due to mainstreaming services, when services really need to be attuned to the specific places, cultures, lifestyles and health issues of the desert population.

The forces that make governments remote from desert regions will never go away. There will always be more people, more voters, more genuine need for services and sales in cities, and so there will always be more attention paid to these places. Desert dwellers need to accept this, just as they accept the fact that droughts and floods are out of their control. The contemporary challenge is to manage *for* this reality, not fight *against* it. Modern technology and regionalised governance structures now offer tools for dealing with distance, remoteness and sparse population or resources, as we shall explore.

This is a vital issue for the world of the twenty-first century. In 2007, for the first time in history, more than half of the world's population was living in cities. Although cities have always wielded power over outlying regions, that power is now consolidated by a growing majority. Cities will continue to grow, squeezing an ever-increasing proportion of people into a tiny area and redefining more and more of the world's lands and seas as 'remote'. At the same time, the megacities rely on these outlands for water, food, energy, clean air, carbon storage, waste disposal and simple escape. How desert regions respond to the challenge of bridging the divide today will help all non-urban areas tomorrow.

3.5 Cultural differences and local knowledge

Remoteness and a relative lack of formal research demand innovation and local knowledge among those who dwell in the desert. In fact, the quintessential image of the outback Aussie battler is based on the perception that people in remote areas persist against the odds and come up with new ways of coping with the environment. This is true whether people are thinking of white settler pastoralists weathering a drought, or Aboriginal 'bush mechanics' keeping their cars going against all the odds, as the television program made famous.[66] Although handouts and an expectation that someone else will provide the services may undermine self-reliance at times, we shall see that there is plenty of evidence of the reality of this image as well. In fact, around the world, local desert knowledge and traditions tend to create solutions that stand in contrast to the failures of recent impositions by centralised bureaucracies. This is true in architecture – local building designs in Saharan towns neatly channel wind-blown sand away from their doors, while modern box houses based on city designs are overwhelmed by dunes.[67] It is true in water management – ancient covered waterways (qanats) and the associated community controls protect the sustainability of the supply, whereas modern pipelines and privatisation result in overuse and salinity. It is true in local peer group institutions that protect land from overgrazing, while commercial ranches set up by government policies may be eroding the land away. There is so much diversity in vast desert regions that a distant central government cannot know all the subtleties, and it is essential to capitalise on the local knowledge of its sparse population.

Remoteness has also made outback Australia the last stronghold of Aboriginal culture – a storehouse of human knowledge and experience in this continent that extends back tens of thousands of years. This is a human endowment unmatched in the world, and it is so close to being lost forever. Two centuries ago, Aboriginal Australia claimed some 700 dialects in 260 languages, which differ from one another as much as Arabic from English. Today all but 80 are extinct. Only 18 were still classified as 'strong' (i.e. being transmitted between generations) in the 2005 national assessment,[68] and all of these occur in remote areas (Figure 7). Sadly, not all remote languages now survive and, when languages are lost, so are immense stores of traditional knowledge (see Chapter 4); however, many of those remaining are desert languages.

Given that Aboriginal society has persisted particularly in the desert, it is not surprising that this might lead to some cultural differences to the norms of city Australia. This is particularly apparent in ways of organising society, decision making, and, most fundamentally, perceptions of the relationship between land and people, as we will see in the next

Chapter. Some of these elements were probably a direct response to the unpredictability of the Australian environment, and it seems that other people who are attracted to life in the desert share some of these characteristics.

In fact, there is growing evidence that social interactions in sparse populations subjected to high levels of unpredictability may be different from those expected in cities, and that this leads to the development of different types of institutions in desert environments, regardless of whether they are Aboriginal or not. Analyst Ryan McAllister has used network theory – an understanding of how people network goodwill, information and the exchange of goods – to show that it is sensible for people to maintain more links in variable environments. Practical examples of this can be seen in networks among pastoralists to look after stock in dry times, and in the intense lacework of mutual obligations among family members that Aboriginal people see as normal.[69] We will see more examples in later chapters.

Considering how different life is for people in desert Australia compared with cities, this is hardly surprising. Dogged pastoralists spread thinly across the vast landscapes, single-purpose townships such as the tourist town of Yulara, mining workforces that fly in to a temporary encampment for a few weeks, and Aboriginal communities that move around between settlements – these are all models of settlement that are alien to cities, and which have customs and cultures and institutions that attract some personalities more than others.

Thus the causal links that drive deserts and their people seem to reach into culture, institutions and even personalities. As climate change drives more variable and unpredictable weather, and social change leads to a growing disconnect between rural and urban populations in many areas of the world, our understanding of these linkages will have increasing universal significance.

3.6 Systemic variability, social uncertainty

Variability and unpredictability is expressed in many other walks of desert life as a consequence of the desert drivers we have discussed so far. Distant markets mean that desert enterprises have little control over the variability in prices. Small populations mean that there is often no replacement for the skills of individuals if they leave. Population mobility can mean that there are big fluctuations in the numbers of people needing services in particular settlements. Distant policy making means that deserts have little influence on central decisions that may have huge local impacts. All these effects create further variability in desert social systems, which we call *social uncertainty* – caused by social factors in contrast to the uncertainty caused by climate.

One major source of uncertainty arises because most markets are distant for desert enterprises, and what is produced in the desert is generally only a small part of what those markets buy. As a result, the enterprises are subject to prices being set somewhere else. Mining is a key industry, but mineral prices are set globally, even though Australia's production is significant in some commodities. In the early twenty-first century, these prices have been high, but desert Australia has little influence over this. There have been mining booms before, such as in the 1960s when there was a shortage of labour in remote areas. The collapse of resource investments at the end of that boom was a major trigger for Aboriginal unemployment and for the social changes that are still playing out today.[70] More recently, there has again been a shortage of labour for mining in remote regions, this time driven by the growth of China. However, the global financial meltdown of 2008–09 shows there is no certainty that Australian mining will benefit indefinitely, with major mine closures and workforce lay-offs. Desert Australia is on the receiving end of all these price fluctuations.

The same happens in many remote industries. Tourism is more affected in remote areas than on the coast by world travel sentiment and the price of fuel, both of which are out of local control. Similarly, the pastoral industry depends on wool and meat prices set worldwide. Only in a few cases do local desert businesses exert significant control over prices – goods and services for local consumption (to a point) and specialised niches such as Aboriginal Art.

A second major source of unpredictability comes from population mobility and staff turnover in small communities. As noted above, a quarter of desert Australia's population turned over in the five years between the 1996 and 2001 censuses. When people leave a particular job in a city, they usually continue to live in the area, their skills are retained, and, anyway, there is a large pool of similar skills to draw on. In small settlements, though, people often leave town altogether, and there is a small replacement pool, most of whom require significant training in a new position. The availability of skills therefore varies greatly over time. This is particularly important for critical skills such as good leadership – the loss of a single leader can be disastrous for a small community. Costs and time delays in recruitment are therefore often substantial. In addition, intense interpersonal relationships in small communities can lead to strong working alliances easily, but can also result in destructive conflicts. This happens anywhere, but people are less able to evade or dilute these conflicts in the smaller communities that characterise deserts.

These are effects on the supply of labour and skills; mobility can also create great variability in the demand for labour and services. As we shall see (Chapter 7), some mining and Aboriginal settlements are characterised by large fluctuations in resident numbers. Service centres such as Alice Springs lose half their population over Christmas, and gain as much as a third of the population again in high tourism seasons. These fluctuations cause big surges and downturns in takings for small businesses, as well as leading to challenges for 'steady state' models of service delivery for education and health. Longreach hospital on the western Queensland tourist trail, for example, has a third of its beds unused over summer, but must have them ready to cope with extra patients from the passing tourist trade in winter.

All of these aspects make for a great deal of unpredictability in labour-related issues. More uncertainty is added by people being unprepared for the difficult conditions. Stories abound of teachers or nurses arriving for a new job in a remote settlement and departing on the same mail plane 30 minutes later – although not frequent, these stories are real! Organisations such as the Centre for Remote Health and the Council of Remote Area Nurses of Australia work hard to prepare remote area professionals for their experiences, but can only partially reduce these effects.[71]

A third general source of local uncertainty is caused by centralised decision making, particularly in government and policy, but also in central offices of companies and other organisations. The dramatic 'state of emergency' declared by the Australian Prime Minister in June 2007 is but one extreme example of this (see Section 4.3). There was no forewarning or knowledge of the resulting 'intervention' in regional government offices, let alone in the remote communities to be affected.[72] Not only was the policy implemented without warning at a local level, it was also designed with little sensitivity to local needs and follow-up.

This is an example (albeit exceptional) of an effect that is played out daily in more mundane ways, as decisions are made by state and national governments that suit the majority of Australians who live outside the desert, yet have awkward implications for desert dwellers. Such impacts are usually unintended, it must be said: taking the form of water quality and building standards or natural resource management and health programs, but most consultation about these naturally occurs where most people live. We will look at some examples of these in later chapters, as well as ways to get around their consequences.

The sources of social uncertainty can build on one another so that well-intentioned policy initiatives fail as a result of staff turnover, or recruitment and training problems, which are compounded by similar failures in related programs and supporting small businesses (see Box 6).

From the point of view of people, businesses, institutions or even governments operating in desert Australia, social uncertainty shares with the climate the characteristic of being *out of any local control*. Social uncertainty makes it difficult for businesses to operate using conventional approaches, hard for organisations to succeed on normal funding models, tough for people to live so far from family and city friends, expensive to live and difficult to access services. A consequence is a self-reinforcing problem: making life harder for people who want to live in the desert, and less attractive for those who do not.

However, it is only a problem if we look at it through the mental models of densely settled, urban areas. Local people cannot manage the uncertainty out of existence; instead, they must re-orientate themselves to manage *for* the variability. For every issue that looks awkward from an urban perspective, there is a desert knowledge opportunity waiting to be grasped. Much of this book is about ways to break the sense that these issues are all problems, and how advantage can be found in the opportunities thus created.

3.7 Responding to scarce and uncertain resources

Clearly it is not only desert plants, animals and pastoralists who are affected by scarce and uncertain resources. In fact, it is desert businesses, desert settlements, desert institutions, even the nature of desert people and their culture. Given this web of causal links, one may well ask what biological 'desert knowledge' has evolved for organisms to survive in the desert? Unpredictable climatic extremes and scarce resources drive all life in deserts and lie behind the other steps in the causal chain (see Figure 4). Plants and animals have adapted to deal with them in a variety of brilliant and cunning ways, which hold many lessons for humans and have application to the way we live our lives and use resources in an overcrowded twenty-first century.

There is a universe of wonderful and intriguing detail in the lives of desert plants and animals, but desert species have five basic strategies for survival.[73] We have already mentioned many of these, but it is time to formalise them, because we shall return to them often in the book. As noted in Chapter 1, to cope with unpredictable resource scarcity the strategies broadly depend on persisting or escaping, in time or in space (Plates 1–3). They are:

- **Persistent locals**: the classic response of many trees, larger mammals and birds, termites and

BOX 6: COMPOUNDING SOCIAL UNCERTAINTY

Peter Ryan was a government officer working on an innovative, community-led law and order program in the remote settlement of Ali Curung, between Tennant Creek and Alice Springs in the Northern Territory. Established under the Northern Territory's Aboriginal Law and Justice Strategy in 1996, the Ali Curung Law and Order Plan enabled the community to have a recognised partnership role in law and justice issues in their region. Significant reductions in violence and crime resulted.

However, the strategy was ultimately vested in distant government departments and, as a result, depended on transient employees from outside the region. Despite the best efforts of many involved, the resulting components of social uncertainty interacted to undermine the local success of the program, as Ryan recounts:[74]

> 'Other factors include poor quality field staff and program managers, and high staff turnover. For example, within 12 months of signing the Ali-Curung Law and Order Plan, several government agencies had a 300% turnover of staff servicing the community. One agency responsible for providing a key program under the agreement had difficulty recruiting for a particular position, which remained vacant for more than nine months. Two other programs critical to the Ali Curung Law and Order Plan were relocated from Tennant Creek to Alice Springs where Barkly region communities such as Ali-Curung could not access them.'

Thus staff turnover (most leaving the region), coupled with the simultaneous failure of a key related program due to recruitment problems, as well as distant decisions to move the base for other critical programs out of the region, compounded to destroy the success of the Plan.

This type of problem cannot be countered by the best of intentions of government to do better from afar: it requires the granting of greater autonomy in decision making and resource allocation to the region, so it can work out its own local solutions – an issue explored further in Chapter 9.

ants is to stay active during the poor times by investing in resource-harvesting infrastructure when times are good – extensive roots for mulga trees, knowledge about large home ranges for dingos, fat stores for antechinuses, or complex social colonies with their food collection structures for termites.

- **Refuge dwellers**: some plants and animals are active all the time, but only because they confine themselves to resource rich patches in the landscape. River red gums grow where there is a permanent watertable, fish depend on mound springs or other permanent water and cycads cling to the shady sides of gorges.

- **Ephemerals**: these plants and animals escape in time. They grow fast when times are good

and then invest in seeds, spores or eggs that can wait out the bad. This strategy is pursued by winter wildflowers, shield shrimps and locusts – the last being an 'irruptive species', like some desert rodents that are at low numbers most of the time, but occasionally explode into plagues.

- **Nomads**: these creatures escape in space but stay within the desert, moving from one burst of rain driven-growth to another, such as budgerigars, chats and some desert waterfowl.

- **Exploiters**: these animals are also mobile, but their strategy is to spend the tough times outside the desert and only come in opportunistically to exploit good times. Waterbirds that come to breed on Lake Eyre from elsewhere in Australia, or the world, pursue this strategy.

Many species combine two or more of these strategies. The most persistent trees produce seeds as survival insurance against a drought so severe that even deep roots are no help. Likewise, many ephemeral plants can behave like short-lived perennials (Plate 8d): if the good rains keep on coming, they keep on growing, but still with the principal goal of producing seed rather than roots to survive the eventual drought. Large animals such as emus try to persist in a particular home range, but, if things get really grim, they up-stumps and become nomadic.

Thus, there are endless permutations and combinations of these individual strategies, but there are also two different, important ways in which multiple species interact to create collective strategies.

- **Dependents**: many species benefit when another gives them a helping hand, accidentally or otherwise. By persisting, locally or in a refuge, the facilitating organism becomes a home or food source for another creature that then does not need to cope with so much variability itself. Examples are: river red gums or mulga that provide sap for many insects, which then themselves become prey for other dependent animals; ticks that live on kangaroos and sleepy lizards; and mistletoes that parasitise all sorts of long-lived shrubs and trees. Some plants also act as 'nurse plants' to others by providing them with shade in which they may germinate. This sort of relationship is unusually common in deserts worldwide, compared with other environments.[75]

- **Self-organising communities**: when resources are sparse, some species find ways of concentrating them. Plants do not do this consciously, of course, but there are ways in which groups of plants commonly 'self-organise'. By chance, they become arranged so that they capture more water and nutrients as these resources flow across the landscape, and

this chance arrangement becomes self-reinforcing (Box 5). These groves harbour a wide range of plants and animals, which, in turn, benefit other organisms living in their vegetation or in the soil below, which may help to slow the water or create macropores in the soil through which the water is absorbed faster.

These two ways in which desert species interact illustrate the crucial ways in which partnerships can help when resources are scarce and variable (Plate 4). Firstly, once key species have discovered how to persist through the extreme variability, they can help to insulate other organisms from the uncertainties. Secondly, the ability to capture and control resources such as water flowing across the landscape enables some species to exist when it would be impossible if the same amount of water was smeared evenly over the landscape (Box 5).

However, all these strategies have weak points. For example, persistent and ephemeral strategists both need to build up reserves during good times – for persistent plants, in the form of roots that explore a large space for water, or big storage organs such as the lignotubers of mallees; for persistent animals, through investment in foraging and storage of grass as for termites; and for ephemerals, in the form of seeds for plants or resistant eggs for insects. Building these reserves is a weak point – if it is interrupted, then the organisms will fail to survive the next dry time. For the refuge-dwelling species, the weak point is damage to their resource rich refuge – pumping out a drought refuge waterhole can cause local extinction of fish, for example. For dependent species, the loss of their protector is a weak point, as when river red gums are cleared or burned along a river and a whole cascade of species are killed with them. Some weak points occur at a larger scale. For nomads such as waterbirds that move from water to water, a gradual removal of individual wetlands may eventually mean there is too far to fly between safe sites for the nomadic population to persist, even though some of the wetlands may still be in good condition. Some of the weak points are more subtle (see Box 7), but they are

BOX 7: THE WEAK POINTS OF DESERT LIFE HISTORY STRATEGIES

The seven strategies of desert organisms have natural weak points (WP). There are four main ways to vandalise the survival of species locally:[76]

WP1: Interrupting investment in reserves for bad times – preventing persistent species and ephemerals from growing roots or setting seeds that are needed to get them through a dry time (e.g. by grazing them as they recover from drought)

WP2: Damaging the resource-rich niches – destroying the special refuges (such as river floodplains, waterholes or groundwater tables) on which refuge dwelling species depend

WP3: Loss of key self-organising species or species for dependents – by removing or weakening certain critical persistent species on which others depend (e.g. by killing river red gums or overgrazing bluebushes)

WP4: Direct damage to self-organised resource flow management – by physically damaging or disrupting the natural flows of water and nutrients across the landscape (e.g. by carving a road through mulga groves and disrupting the normal drainage patterns)

There are also four larger-scale weak-points that are triggered when local human activity is accumulated over a whole desert region or interacts with changing climate. This different level of vandalism – which is harder to detect – is like 'death by a thousand cuts':

WP5: Degradation of 'pulse networks' – nomads depend on there being a network of places where resource pulses may occur; if that network is thinned by losing individual locations, or the locations become disconnected, at some point this network quite suddenly stops working

WP6: Decline in resources – persistent plants and animals live 'close to the edge' by harvesting sparse resources that others cannot; relatively small declines in the availability of these resources can strain their resource harvesting capabilities too far, whether this is caused by climate change (e.g. slight drying) or management (e.g. moderate but uniform grazing)

WP7: Extended periods of low resources – both ephemeral and persistent local organisms are susceptible to longer dry spells, either through climate change or through human activity, which alters the return time of pulses (e.g. water extraction from ephemeral rivers or groundwater).

WP8: Changes in the timing of resource availability – for nomads, there must not only be a network of locations where pulses of resources occur, but these must also occur at different times so that there is always a rich site somewhere. Degradation and climate change may affect this synchronicity.

all identifiable. The key issue is that humans often trigger the weak points as we exploit the landscape, sometimes with disastrous consequences for desert plants and animals (Plate 5).

Understanding the weak points in natural ecosystems is important for people who live in deserts and are trying to manage their natural resources sustainably (see Chapter 5). But these strengths and weaknesses also apply to human life in the deserts overall: businesses, settlements, services and governance structures also have to manage with scarce resources subject to climatic and social uncertainty (Plate 6). We will draw vital insights from the opportunism and persistence of nature in later chapters.

3.8 Living with the desert drivers

Three-quarters of Australia, and at least a third of the world's land area, is subject to desert drivers – particularly uncertainty, scarce resources, sparse population and remoteness – and so face quite different issues from the places where most policies are formed and most decisions are taken. Living in, and managing, desert lands is conceptually a world away from inhabiting a neatly ordered and regular European landscape or Australian coastal cityscape.

When settlers from Europe first encountered Australia, Aboriginal people knew very well how to cope with the continent's vagaries. But they did not know how the colonisers' livestock, water use and way of life would interact with the continent; very few of the settlers made a point of asking them, or of studying how and why the locals did things. As a result, there was a struggle over the 'best lands' – often the patches of richness that maintain and recharge life in the continent – and many of these were damaged or lost.

Consider the example of grazing. Europeans came from lands where an extreme dry year – say one year in thirty (a managing lifetime) – implied getting less than two-thirds of the average annual rainfall. In one year in thirty, Alice Springs receives less a third of its average rainfall; in fact, central Australia gets less than two-thirds of its average nearly one year in three![77] For people from such a stable European environment, it was natural to regard a dry year as unusual: herders decided on a stocking rate for sheep or cattle based on a normal year and stuck with it, perhaps just selling a few extra in a dry year. It soon became clear this formula did not work in Australia – early pastoralists had to find other approaches.

Adjusting livestock numbers up and down by 30 per cent or more every second or third year to suit the variations in rainfall is basically impossible, so inevitably there were far too many animals left on the land when droughts began. As Baldwin Spencer remarked about the state of common knowledge in 1886, 'These rapid changes have, however, led to ruinous losses among the pastoralists, as people with a meagre knowledge of the climate … when the inevitable drought occurs … they find their stock dying by hundreds of thousands …'[78]

As a result, livestock ate the perennial plant reserves (weak point WP1 – see Box 7), grazing out the best areas or refuges (WP2), sometimes destroying the keystone species (WP3), and trampling channels through self-organised flow systems (WP4). Many of these effects happened over wide areas, triggering weak points WP5 and WP6 as well. The only saving grace was that the lack of surface water in droughts excluded the stock from many parts, thus limiting the damaged area of country. However, as bores were introduced for water, the areas of untouched country shrank.

Today, we understand pastoral management far better, but it has taken 100 years for us to do so. And we are now beginning to appreciate that it is not only native plants and animals that must deal with the desert drivers. As shown in this chapter, the effects of uncertainty, scarce resources, sparse population and remoteness affect the whole human enterprise of desert living. They affect businesses, services, settlements, governance and indeed people, in every aspect of their lives and personal feelings.

All desert dwellers have to learn to exploit these drivers – to take advantage of the shifting patterns of richness and deprivation which they provide. The Aboriginal people who inhabit the deserts did not originally go there because they loved deserts. When they arrived 40 000 years or more ago, these regions were bountiful with water and life; the immigrants had to learn the hard way how to survive when things dried out due to natural climate change.

Today, we face the strong probability that things will continue to dry further in most of the world's deserts – and in many of its food bowls and savannah regions – and we too must learn how to adapt as things dry out, this time as a result of self-imposed climate change. The question is, how do we turn this desert knowledge to the benefit of those who now inhabit deserts, as well as those soon to do so through climate change and desertification?

The central aim of this book is to explore how to create sustainable livelihoods for people in deserts that support them economically, socially and culturally, and to enable them to continue to live there and look after their landscapes. From the success of this enterprise emerges understanding for the future in many other parts of the world.

4

Desert survivors

'Thus there will be one area which belongs to a group of men who call themselves kangaroo men, another belonging to emu men, another to hakea flower men, and so on, almost every animal and plant which is found in the country having its representative among the human inhabitants.'

SPENCER AND GILLEN 1899[79]

So wrote Baldwin Spencer after his first travels through central Australia in 1884, highlighting how every component of the ecology and environment of desert Australia was intimately woven into Aboriginal life, and how Aboriginal life was an essential part of that environment.

It is impossible to tell a tale of desert living in Australia without talking of Aboriginal peoples, for reasons happy and sad, inspirational and shameful. A 40 000 year history of adapting to the Australian environment is cause for wonder. The persistence of ancient knowledge into the modern age is a potential source of insight. The rapid loss of languages, and with them an enormous body of practical knowledge about how to live – and live well – in Australia, is sorrowful. The state of health and wellbeing of many Aboriginal people within this 'lucky country' is scandalous, but it also provides an insight into the general failures of desert governance. And the future of desert Australia is inseparable from the future of the quarter of its inhabitants who are Aboriginal. In this chapter, we explore these issues briefly because Aboriginal heritage is a key uniqueness of desert

Australia, and it is no empty rhetoric to say that this land's best option is a shared future.

Author and cameleer, Robyn Davidson, has spent time with nomads around the world. She points out that most people today did not grow up as nomads and so cannot see into the deep crevices of another way of life based on 'differences in the very foundations of reality'.[80] Thus we approach the rest of this chapter in the spirit of exploring publicly known aspects of Aboriginal culture, which seem from discussions with many people to hold strong messages and lessons for our common future – whether or not we understand them fully.

4.1 Once were wanderers

As we saw in Chapter 2, Aboriginal ancestors have been living in inland Australia for at least 40 000 years, and have found ways of living with the environment through cycles of climate change. As we have noted, they were not originally drawn to deserts by a love of harsh dry places – when they first went there, central Australia was filled with great waters, alive with fish and wild game, and covered with trees. These facts exemplify the changeability of our continent. However, for several millennia now, Aboriginal desert dwellers have been coping with scarce resources and variable rainfall, which combined to support only a very sparse population. While there might have been a person to every two to twenty square kilometres on the coasts of the Top End or Cape York, there was only a tenth to a twentieth of this in the Western Desert.[81]

This is not surprising. If you live in an environment with sparse and unreliable resources, then you have to be able to harvest those resources from a very large area. This was what Aboriginal tribes in central Australia did. By contrast, Yolngu people in the Top End accessed much more reliable resources around the now flooded Lake Carpentaria and the coastline of Arnhem Land, and were able to spend most of their time in much smaller areas. The area that each Yolngu group called their own was only around eighty square kilometres, while for Pitjantjatjara people around Uluṟu it was more like 3000 square kilometres. In south-eastern Australia, there were semi-permanent Aboriginal settlements associated with stone fish traps in the Darling/Barwon river near Brewarrina in New South Wales, and eel traps in western Victoria; these provided much more reliable concentrations of resources than central Australian Aboriginal tribes could have ever expected.

So, the desert groups had to be able to range over large areas. Western Desert tribes were highly mobile at various scales. Family groups identified with a region that included a few reliable waters. In good times, people spread out: resting these key refuge areas and harvesting resources very widely. In drier times, they would fall back to the permanent waters. But the Western desert environment is not slightly variable: it is utterly fickle. From time to time, even mobility within a large range was not enough. During very severe droughts, bands moved out of their own country altogether for a while. They depended on their social connections with neighbouring groups and verbally transmitted knowledge from earlier generations to survive elsewhere.

Maintaining these relationships meant meeting up at times of plenty, intermarrying and trading goods and lore to create the goodwill and social capital that everyone can draw on when times are tough. Elders in each group were especially important, because they were the sources of this essential survival knowledge from generations gone by – a striking contrast with modern western societies, which have a tendency to undervalue older people as soon as their day-to-day knowledge appears dated.

Everything needed for life came from the environment. There were people who were specialists at making particular tools – and men and women's roles were more or less differentiated. But, in a hunter-gatherer society, everyone has to know and do everything needed for survival reasonably well. Spencer wrote 'Councils of the elder men are held day by day … all the traditions of the tribe are repeated and discussed, and it is by means of meetings such as this, that a knowledge of the unwritten history of the tribe and of its leading members is passed on from generation to generation'. It is extraordinary to reflect on their depth of knowledge about hundreds of plant and animal species, thousands of square kilometres of geography and geomorphology, and a hugely complex array of social relationships.

These diverse types of knowledge were passed down relatively flawlessly through innumerable generations by word of mouth – assisted by songs, rituals and experiences. The sum of this knowledge about the animated landscape was timeless, so that stories mix up what westerners would call past, present and future. The Ngurunderi dreaming of the Ngarrindjeri people of South Australia recalls Kangaroo Island being separated from the mainland by sea level rise 9500 years ago. Equally old stories from the Gulf of Carpentaria carry the Lardil people's memories of sea level rise as Lake Carpentaria flooded at rates that must have meant the shoreline moving at fifty to a hundred metres per year at its peak.[82] These are orally transmitted memories that date back 10 000 years! Much as we may think of mass-produced books and computer disks as being permanent records, there are few media in the world that have lasted a thousand years, let alone 10 000. Even the great pyramids of Egypt or tombs of Sumer are barely half as old.

These early Australians were also pioneers in the rise of human self-awareness, expressed in works of art possibly dating back 30–40 000 years – as old, or older, than those so far found anywhere on Earth.[83] Europeans are accustomed to thinking of the cave paintings at Lascaux in France as most ancient art, yet these are no more than 17 000 years old. In fact, other art in Europe may be as much as 32 000 years old, but Aboriginal people were probably exercising this form of expression at least 10 000 years earlier.

It seems inevitable that people who live close to their natural environment do not see themselves as separated from it; all hunter-gatherer societies are like this. Humans, animals, plants, land, wind and fire are all interdependent and inseparable, as is obvious in every daily, hourly action – collecting food, using fire, making shelter or finding water. People depend on what the natural environment provides, and what the environment provides is affected by what people do. They see and experience the feedback every moment of their lives.

Although the desert environment provides, it is also harsh. Small mistakes are often fatal, so people must work together. Because resources are scarce, there cannot be too many people in one band. The result is small, close-knit groups, with strong rules and a powerful sense of obligation to each other. In fact, the harsher the environment, the more this imposes a need for rules. Over time, those groups that survive most successfully will have the strictest sense of rules, the strongest culture and the most powerful reciprocal obligations. They may also be more egalitarian and flexible: exclusive rights to resources, and consequent hierarchies, only seem to evolve where resources are more plentiful.[84]

Rosie Kunoth-Monks from Utopia, 250 kilometres north-east of Alice Springs, says,[85]

'The old people had the ability to read the environment and know when shortage of food or water or some other life giving force was imminent. They would tell us something was in

short supply and we had to tighten our belts ... We were instructed not to hurt or take certain species until they had regenerated. In these times there was a reason for discipline, skin relations, sacred songs and performance of ritual because it strengthened survival.'

There were (and still are) complex interactions among these desert groups. Linguist and anthropologist Carl Strehlow remarked on how many of his informants in the 1890s had personal knowledge and experience of extraordinarily large areas of land. Young men would often travel with other groups, learning the country and the ways of surrounding areas, up to hundreds of kilometres away – but most of them returning to the family group eventually. Such movements would often lead to intermarriage between different groups, creating massive networks of blood connections that could be called upon during hard times. The connections were further enhanced by complex exchanges of sacred rituals and objects between regions – noticeably more so in the desert than in richer environments. Major ceremonial gatherings probably only occurred every decade or two during prolonged good times, but then these could take weeks to prepare for and hold, with peoples arriving from many directions and distances with well-defined roles to play. These linkages, exchanges and learning experiences were an investment in the group's ability to move greater distances in order to survive when the need arose.

They also provided opportunities to pass on vital knowledge. Bruce Chatwin's book *The Songlines* delights in listing the extraordinary diversity of expression by which pre-literate cultures around the world recorded information and geography in spoken memories.[86] Often these weave memorable stories that incorporate mnemonic details for survival. In his memoires, stockman Walter Smith recounted how, as part of his initiation teaching, he had gone with Sandhill Bob – his tribal guide and teacher – deep into the Simpson Desert in 1924. His companion had not visited the area for many years and was singing the songs of the country to help direct them. One day, he was a 'little bit lost' when trying to find a soakage after travelling many kilometres over 'broken channels and hummocks of the floodout country, where features were altered with every major flood and every major dust storm'.[87] After some searching, they found the stones that marked the soakage, hidden by some long grass – afternoon shadows had thrown Sandhill Bob out by a few paces over a whole day's travel in country most people would have thought featureless. Memories sparked through songs taught to him when he was young carried the precise and reliable instructions needed to save a family from dying of thirst. Remarkably, they also demonstrated that an oral tradition thousands of years old was capable of achieving accuracy as good as the ubiquitous on-board GPS navigation tool of the satellite age; it was just as effective but dependent on rather more expertise.

Fiona Walsh, who worked with Martu people in Western Australia, observed the remarkable change that older women underwent when they were on their country. In meetings and in town they would be quiet and withdrawn, but travelling the land they knew so intimately they would become active and confident: calling to the characters associated with the site, introducing themselves and any visitors, and asking for support with hunting.[88] Recording tales from Walpiri elder Darby, Liam Campbell recalls the same animation of people when travelling through country.[89] They would have stories and memories about the plants and animals and landscapes and activities that bound them to that land. The rehearsal and retelling of these stories is an incredibly powerful storage medium – as well as an enduring memory of the changing nature of the Australian continent, which recent generations have not yet learned to value for what it is.

Small groups without written records also need effective mnemonics to know where each person stands in relation to everyone else. Aboriginal groups have diverse, but widespread, kin systems that help track people's relationships to each other, and which include complex avoidance behaviours. In small groups this helps to avoid inbreeding and to ensure the continual renewal of the kin networks needed for

survival. The systems varied from four to as many as 16 skin groups, and were still spreading and evolving at the time white people arrived. At the least, mothers-in-law and sons-in-law were placed in different groups that generally could not meet. Depending on the region, marriage was often tightly prescribed in ways that made strong links between neighbouring groups. However, arrangements were diverse, and could be flexible where small group size meant that the preferred rules could not be followed.

In the past, and still today, skin names in some regions also conveyed the person's sense of identity and belonging with respect to totems – particular resources or locations. In the desert, such linkages were usually associated with the land and its food and water resources, creating a sophisticated way of sharing, organising and storing ecological knowledge. Above all, there were usually at least two people responsible for any one totem: one who might be loosely translated into English as an owner or custodian, and another as a manager or caretaker. Both would need to agree about the use and care of this particular resource (see Box 8) – providing a much greater insurance against loss of knowledge if something went wrong

It has become increasingly clear over recent decades that Aboriginal people really managed their country actively. As economic anthropologist Ian Keen says, they should be talked of 'hunter-gatherer-cultivators', not just hunter-gatherers. In desert country, the most obvious example is 'firestick farming': the pervasive management of resources and country with fire – whether to burn and regenerate or burn around and protect. Fire was used for reasons such as signalling and just making country easier to walk through – activities that might be confused with explicit management (because everything is linked). However, Fiona Walsh records Martu people actively protecting certain stands of mulga, and central Australian botanist Peter Latz, who grew up at Hermannsburg, talks of sacred areas around which fires are lit to prevent accidental burning of the sacred site.[90] There are many other recorded examples of such deliberate uses of fire. Another form of management was protection from overuse. There is

BOX 8: RESPONSIBILITIES IN THE WESTERN DESERT

Desert Aboriginal groups had sophisticated relationships to country and other totems, which promoted resilience in their systems of knowledge and care.[91] The Warlpiri people of the Tanami Desert, for example, inherit their country through their father's line, such that men's children in this line are known as kirda ('owners' or custodians) and the women's children are kurdungurlu ('managers' or caretakers). These links mean that men and women, kirda and kurdungurlu, have complementary rights and responsibilities with respect to land – each needing to consult with the other before action can be taken in general. The rights include 'speaking for this country' and controlling other people's access and activities on country, including burning. In effect, two sets of people ensure that each does the right thing by the land, and a wide range of people ensure that the knowledge and responsibility is passed on.

Such a system ensures resilience in looking after such an uncertain and variable land and its resources, when one responsible person might die at any time, or might misuse the resource to the disadvantage of the group as a whole. This type of thinking might provide a better model for leasehold land in remote areas of Australia. At present, the local leaseholder is the manager and a distant government is the owner. A better arrangement might be the leaseholder as manager, but an independent set of local resource management interests as custodians.

plenty of evidence in central Australia of core waterholes and other areas being regarded as sacred so that no plants or animals were harvested there, thus creating a refuge. Dame Mary Gilmour cites a series of similar examples from around Wagga Wagga told to her by her early settler father.[92]

Considering they moved on foot across apparently harsh country, the evidence for long distance Aboriginal trading routes is extraordinary. Marked out by songlines of stories exchanged in ceremonies along the route, valued materials – such as ochre for painting and pituri for chewing, kangaroo skins from the south and pearl shells from the north, axe heads and grinding stones – were all traded across thousands of kilometres, along with songs, dance and art. Dreaming stories laced the continent over extraordinary distances, from south-western Australia to Cape York, and South Australia to the Kimberley (Box 9). Taken together, all the stories and the trade routes they describe are like prehistoric forerunners of the internet – a dynamic web exchanging knowledge, culture, ideas, beliefs and human relationships. Trade in ochre and pituri extended over at least 1.4 million square kilometres around the central desert, with distances up to 3000 kilometres involved. Indeed, writing in 2000, archaeologist

Isabel McBryde comments that songlines supported exchange networks 'measured in thousands of kilometres. They are among the world's most extensive systems of human communication'.[93]

Many of these customs and practices of desert Aboriginal peoples observe the principles of the various strategies for desert survival explained in Chapter 3. Although we normally think of the traditional desert living style as nomadic, in fact, desert people more often behaved as 'persistent locals' with large home ranges most of the time. Only occasionally, under really extreme conditions (as well as for trade and marriage at other times), did they become truly nomadic and move to other regions. Humans, above all else, are adaptable, and Aboriginal strategies undoubtedly also contained an element of refuge dwelling, at least in poor times.

There seems to have been a general sequence of events between times of bounty and hardship. In good times, when ephemeral playa lakes and other water sources made widespread movement easy, the people used normally inaccessible areas within their general home range. As times became drier, people retreated to the better lands in their home range where there was more reliable water and food. Then, in periods of extreme hardship, they might

BOX 9: SONGLINES AND DREAMINGS

The English term 'dreamtime' seems to have been coined in 1899 by Baldwin Spencer and Frank Gillen to translate the Arrente word *alcheringa*, now more commonly rendered as the Dreaming (as is the more widely known Pitjantjatjara word *tjurkurpa*). The concept, almost universal among the many different Aboriginal tribes, encompasses timeless creation stories about ancestral beings, but also the patterns of life and culture for Aboriginal people. The stories about ancestral beings include a great deal of instruction about how to behave, but also about journeys through the land originally taken by the ancestors. These travel stories are the songlines – 'the labyrinth of invisible pathways which meander all over Australia'.[94]

There were many great Dreaming routes across the continent, passing through the lands of many different tribes. These were linked by a series of songlines, or story strings, that changed subtly from one tribe's lands to the next. Sacred knowledge flowed along the route through these songlines in the course of trade and ceremonies. Researcher Dale Kerwin has analysed the relationship between many of these and known trading routes.[95]

For example, the 'Native Cat Dreaming' follows the travels of spirits from the Port Augusta region in South Australia right across the country to the Gulf of Carpentaria. It followed one of the major trading routes for pituri (leaves of the shrub *Duboisia hopwoodii*, which contain nicotine and were chewed for their narcotic effect), intersecting the 'Fire Dreaming' which connected peoples on the west and east of the Simpson Desert. The public part of this story follows an ancestral fire striker, which is lured from one tribe's lands to another by two men, who escape on ancestral snakes. The fire leaves memorable marks on trees and rocks as it travels.

These stories, in turn, were part of the extraordinary geographic extent of the 'Dingo (or Two Dog) Dreaming'. This large set of songlines links south-west Australia through central Australia to Cape York, the Kimberley and South Australia – also particularly associated with pituri trading. Some stories were perhaps more about knowing local resources, such as the 'Two Boys Dreaming', parts of which identify sacred places where only men may go, or waterholes where swimming is not allowed, as well as landmarks identified with the boys.

move out of the home range altogether, calling on relationships with surrounding peoples to obtain access to other lands until conditions improved.

Perversely, given an intimate knowledge of how and when to access scarce and variable desert resources and of how to develop reserves against the worst times, variability means the desert can also be a bountiful place for small groups. Anthropologist Ted Strehlow describes how the Wangkangurru people inhabited a large area of the Simpson Desert where there were no permanent waters. However, a few soaks were very reliable and this enabled them to range over a vast homeland with a strong sense of security. As Strehlow says, 'had their swamps and lakes dried out in long droughts, they would have had the right to retreat temporarily to the permanent waters situated in the lands of their neighbours, with all of whom they had close social ties through intermarriage and through the sharing of sacred traditions'.[96] Indeed, the Wangkangurru people finally left their homeland of sand dunes and soakages during severe drought in 1901 to go to missions in South Australia: movements of the type they would have made many times before – only this time, they never really returned.[97]

Today, the Simpson Desert is an essentially uninhabited region shared between South Australia, Queensland and the Northern Territory, where the land is no good for cattle. Yet, in another age, with different aspirations and approaches to land management, and with generations of investment in local knowledge, it was the heart of the productive 'grain belt' of pre-European Aboriginal lands – the region where grass seed could be reliably harvested most years, and which lay along the trading routes for ochre, pituri and valued goods from the north, such as pearl shell.

The example of the Wangkangurru illustrates how Aboriginal people obeyed the desert rules for survival – build up reserves in good times that can get you through the bad ones. These reserves are not stores of canned food, though they may include fat reserves on the body, which help survival a little longer into a dry time. But, importantly, they are 'reserves' of knowledge about country, which allow access to its full range of resources. They are also plant and animal reserves in core refuge areas that build up thanks to protection in good times. They are also social reserves through connections to neighbours, which allow access to other country as a last resort. As with plants and animals, building these reserves means trade-offs – spending time to learn and travel in exchange for knowledge, and spending time and energy and gifts (including marriage) to build up reliable social capital. In good times, desert people could not just sit down and rest; they had to take the opportunity to invest in ceremonies and to travel in preparation for the next tough time. A moment's reflection will reveal that this is exactly what Australia needs to do in order to weather climate change today – invest in our knowledge of how and where to survive. This is a very different, proactive approach to life compared with assuming that someone will pay you drought relief or social security.

4.2 Wisdom from the desert

It is said 'The last ones to understand the nature of water are fish'. We find it hard to know exactly why we do the things we do in our own individual cultures because they are so much second nature to us. So it helps to distinguish four different, but overlapping, types of knowledge that everyone uses; there are Aboriginal and non-Aboriginal examples of each. These types are: usage knowledge, knowledge about practices, organisational knowledge and worldviews (Box 10).

Aboriginal people had extraordinary *usage* knowledge about their environment. Almost every plant had a use – and knowing when and where a particular flower or fruit or whole plant was going to be available was essential to survival. Early European settlers took advantage of this knowledge, realising the medical benefits of native plants such as ti-tree and eucalyptus oil. Pharmaceutical companies have sought to exploit the knowledge further, in the course of bioprospecting for new chemicals in recent years. Notwithstanding famous successes elsewhere in the world, such as the anti-hunger properties of the San bushmen's *Hoodia gordonii* plant in the Kalahari[98], there have been few notable exploitations of the rich botanical treasure chest of desert Australia. Perhaps the process has a sense of stealing Aboriginal cultural knowledge. As a consequence, the most recent efforts to build a bush products industry are exploring ethical business structures and agreements to preserve for local people the benefits of sharing their knowledge, in much the same way that software developers and drug designers aim to benefit from their innovations. The desert knowledge is real: Aboriginal locals are well aware of how the flavours of common plants such as bush tomatoes and bush bananas vary across their lands, and can pick out the best varieties to breed. This is a genuine opportunity for the future, as the bush foods industry transforms itself from a few million dollars per year to hundreds of millions over the next decade (see Chapter 6).

Aboriginal *knowledge about practices* is also being picked up and recognised in the modern world. Early pastoralists were quick to use Aboriginal knowledge about the land and use of fire to look after their sheep and cattle. Today, the greater interest is in conservation management. Aboriginal knowledge of how native mammals used the landscape and of what sort of management of fire and feral animals is likely to work has become a vital ingredient in helping to conserve endangered species. The opportunity to combine conservation management with maintenance of culture – and obtain the ancillary benefits of healthier lifestyles and good bush tucker – is justifying the establishment of a series of Indigenous Protected Areas (see Chapter 5). With a growing recognition of the importance of maintaining active management

BOX 10: TYPES OF KNOWLEDGE

Aboriginal knowledge is really an integrated whole, but western science tends to engage with different aspects of it in different ways:

1. *Usage knowledge* is like the dictionary definition of the food and medicinal attributes of specific plant resources, or where to find animals, or how much water can be taken from a particular rock hole, or where the best ochre of a particular colour is to be found. From outside Aboriginal culture, this is the easiest knowledge to engage with. In a long history worldwide of ethnobotany, traditional peoples and scientists have together documented knowledge about plants and landscapes.

2. *Knowledge about practices* is more like looking up how to do something in an encyclopaedia – particularly in relation to managing the resources that are identified in the usage knowledge. This includes using fire to clean up country and to promote the growth of bush tomatoes for the next season, for example, as well as cleaning out waterholes to maintain the water supplies for people (and animals) in the area. Such actions have obvious outcomes. However, these practices extend to ceremonies intended to increase animal numbers or bring rain, which may not seem so logical – at least to the outsider. Of course, touching wood or not having floors numbered 13 also have no obvious logical consequences (though not walking under a ladder is a bit more functional!), but such actions are still common in mainstream Australian society. These practices are deeply embedded in belief systems and are part of the fabric of culture, regardless of their functionality.

3. *Organisational knowledge* is about ways of organising people within a group, and the group's relationships with wider society. This type of knowledge can seem very strange and alien when viewed from another culture. Examples in Aboriginal cultures include: the kin system, with its taboos on interactions; powerful senses of reciprocal obligation among extended family members; ways of making decisions in the group; respect for elders; and the whole structured system for initiation and gradually passing on more and more traditional knowledge. Of course, white Australians have equivalent customs, but because they are lived every minute of the day they go unnoticed (Do you hold the other person's eye when you are talking to them or avoid contact? Do you expect your view to be heard in a meeting, or only to listen to the most senior person present? Do you think it is acceptable to 'buy votes' in elections?).

4. *Underlying worldview*, which is common to many societies tied closely to the land, is a fundamental, powerful sense that the environment, the people, their ancestors and the whole body of knowledge associated with life are simply parts of one whole. Such a view contrasts powerfully with the dualistic perception of the separation of humans and their environment that has pervaded European thought for many centuries.

across vast areas of inland Australia that have been vacated by their original inhabitants, there are again enormous opportunities for future livelihoods for Aboriginal people in the continued application of these practices. Of course, like all knowledge, it can be re-contextualised. Traditional practices and learning can help technological management, but the reverse is also true, as a Warlpiri man pointed out: 'now satellites can spot bushfires, [we] can go to the location with proper protection … and teach young people'.[99]

Usage knowledge and knowledge about practices slot relatively easily into the western mindset. It is *organisational knowledge* – how to organise in the face of variability and small populations – that is perhaps the least appreciated and most intriguing. There are many examples, from kinship structures, through relationships to totems and country (Box 8), to day-to-day ways of conducting business. Non-Aboriginal people who work for a while in remote Aboriginal communities soon come to realise that there are Aboriginal ways of holding meetings, of consulting and reaching consensus, and of working out conflicts. More than anything, these take time compared with the rushed meetings of western bureaucrats. Umoona elder George Cooley said at a meeting in Longreach recently, 'You won't know who I am and where I come from in five minutes',[100] referring to the short time he had been allocated on a busy conference agenda.

By central bureaucratic standards, desert ways of doing things may sometimes seem slow and unwieldy, but they invoke and maintain the social networks that are so important in small communities. This book is about coming to understand that life in remote areas runs in different ways from that of large population centres. This means that different models of engagement and governance are needed for desert Australia, and that there are lessons to be learned from Aboriginal ways of organising (see Chapter 9). Some of these are much more similar to the ways in which other remote communities operate, such as those of pastoralists, than either pastoral or Aboriginal ways are to those of cities.

Last, but not at all least, is the *underlying worldview*. Much has been written elsewhere about worldviews that link humans and their life support systems as one. This could be the source of a new land use ethic in Australia: one that is already shared by some who live on the land, but is not yet a universal or deep part of our Australian psyche. To become fully Australian, we need to develop an Australian worldview that is founded upon the very nature of the continent itself. People holding such a worldview would not allow themselves to overuse, pollute or waste precious water, or to damage the fragile soils, squander scarce nutrients or to extinguish the species that comprise the web of life on which we all depend. Such a worldview could become the source of a uniquely cross-cultural Antipodean 'dreaming' that underpins a clearer Australian identity – if only it

can develop and diffuse before too much is lost (see Chapter 10). The lesson learned by Aboriginal ancestors arriving on the continent those many thousands of years ago is that, to live here well, you must understand and live with the nature of the country. This still holds good today. We must all stop thinking like Asians, Europeans or other people from elsewhere – and start thinking like Australians; this is an opportunity for a very special cultural convergence.[101]

We have dwelt on these different types of knowledge because they help to relate what people in Australian mainstream cultures do in their own daily lives to what might otherwise be trivialised as irrelevant cultural oddities. City folk, too, have their *usage knowledge* about which shops to go to, what a light switch does, and which foods are good to eat. They have *knowledge about practices*, such as how to look after the dog, what sort of fuel to put in the car, and how to keep cutlery clean and hygienic. They also have less functional practices, such as washing the car every weekend, choosing ties to wear or keeping the lawn at a particular length. City people have their *organisational knowledge,* too, which is so engrained that they do not think about it, such as which seat to take on the bus, how to address the boss and what rights to expect in their daily lives. And most Australians have grown up within a fundamental *worldview,* which divides 'man' and nature, and which at least partly determines the relationship between our culture and our environment – our infatuation with consumerism, and our obsession with conquering nature.

These different types of knowledge have their role to play in the environments that we live in day to day. Many of them, such as the rule for which side of the road to drive on, we could not get by without. Others are less critical, and some (wearing a suit and tie on a hot Sydney day?) are even dysfunctional, but all contribute to the rich fabric of our culture, and distinguish our culture from others.

So too with Aboriginal traditional knowledge: some elements, if Aboriginal people are prepared to collaborate and share them with us, provide deep pointers for how to manage environments that vary as much in time and space as our Australian deserts do. Other elements will turn out to be vital to Aboriginal cultural fabric, but perhaps not crucial for management today. Yet others may even prove dysfunctional because the world in which this culture finds itself today has changed so dramatically.

Of course, it seems rather perverse to be subdividing knowledge when the key aspect of Aboriginal life is its holistic approach – the fact that a plant was used as a medicine, was managed by burning, contributed to someone's identity or was just part of an integrated world, were all seamlessly interwoven facets of a timeless universe in Aboriginal cosmology. But, it helps those of us with a western upbringing to make sense of the ways in which the modern world could learn from Aboriginal knowledge and come to appreciate what it has to offer us.

The future of deserts must draw on many sources of knowledge, and traditional and contemporary Aboriginal knowledge is only one part of this mix. However, it is the least recognised and most trivialised, so Aboriginal knowledge and ways of doing things will surface in many places in this book as we explore how the world is changing.

4.3 A collision of cultures

On 21 June 2007, in a ten minute press conference, the then Prime Minister, John Howard, and his Minister for Aboriginal Affairs, Mal Brough, announced an emergency federal intervention in the Northern Territory, essentially because there were suspicions of rampant child sex abuse in remote Aboriginal settlements. The trigger for this action was a report by Rex Wild and Pat Turner, titled *Ampe Akelyernemane Meke Mekarle 'Little Children are Sacred': Report of the Northern Territory Board of Inquiry into the Protection of Aboriginal Children from Sexual Abuse.*[102] The response was to rapidly mobilise additional police, doctors and the army and send them into many Aboriginal settlements in the Northern Territory in a show of resolve. On its coat-tails were mooted changes to land tenure, the suspension of human rights, the quarantining of people's income, the removal of the permit system for Aboriginal lands and further

termination of the Community Development and Employment Program – an Indigenous-specific, low paid work program.

Was it truly a 'crisis'? In the sense that there was a growing suite of critical issues that needed action: most certainly. Many commentators, including many Aboriginal leaders, had been calling out for more concerted attention and action for decades.

Was it truly an emergency? In the sense of a sudden unexpected danger such as a tsunami or a bushfire: most certainly not. There had been a stream of government reports over the previous two decades that documented the increasingly dire state of substance abuse, health, education, law and order on remote Aboriginal settlements and in Aboriginal communities in desert towns. Ironically enough, the bibliography of *Little Children are Sacred* itself lists at least a dozen of these (as well as quite a few related to Australian society more broadly).

Was a major public investment warranted? Undoubtedly, but there was little point in this if it involved approaches that undermined any hope of success. As David Ross, director of the Central Land Council, said at the time, any response '… needs to be considered and inclusive … [because] international examples clearly show that, to achieve lasting change, efforts must be made to build indigenous capacity to solve their own problems',[103] whereas, he said, this response was a 'frightening example of centralised control'.

In truth the 'emergency intervention' arose from a confluence of political opportunism and genuine public concern that such things could be happening in this 'our lucky country'. The underlying causes had been a long time coming.

As mentioned in Chapter 3, in 1788 there were probably 260 Indigenous Australian languages, supporting at least 700 dialects. Today, about 180 of the languages have died out completely, mostly with little documentation (though there are brave efforts to revive a few). Of the 80 still spoken, over 60 are severely endangered and are likely to die out altogether by 2030. The remaining 18 are still being handed on to children as a first language. But they are all classified as endangered, and current trends ominously suggest that no Indigenous Australian language will be spoken on a daily basis by 2050.[104]

When the museum in Baghdad was damaged and looted in 2005, there was worldwide outcry. When cultural artefacts marked with UNESCO's white and blue plaques in the old city of Dubrovnik were bombarded in the Serbian war in 1991, there was disgust on the front page of every major newspaper.[105] The burning down of the Great Library of Alexandria is regarded as one of the greatest acts of vandalism in the ancient world. Yet, with every language lost, an incomparable store of traditional knowledge about our continent, how to live in it and ways of thinking inevitably vanishes as well. How can we have allowed this to happen here in civilised, educated Australia, where we take such a pride in our diversity of cultures?

This is not just 'a bit of a shame': along with irrevocable losses of the world's oldest living culture go precious insights about how to live well in Australia, as times, circumstances and climates change around us.

These losses did not happen in a single bombing raid or torching, but insidiously through a mixture of conscious colonisation and casual cultural decay over 200 years. In some ways, desert and northern Australia have been lucky: most of the 18 languages that remain relatively strong have persisted in these regions that were furthest from the colonising influences (Figure 7). Thus Aboriginal and non-Aboriginal desert dwellers together still have an opportunity to maintain a living cultural understanding of traditional insights about deserts and our own country that were accumulated through thousands of years of human experience.

Meanwhile, though, it would be a very rose-coloured view to suggest that all is well in remote Aboriginal communities. Health, education, personal security and housing – all have catastrophic statistics. The mortality rate of Aboriginal people in central Australia is nearly three times that of the total Australian population and, while the rate is improving, it is improving more slowly than for non-Aboriginal people. Heart disease is six times as common for Aboriginal males and eight times as common for Aboriginal females than among non-Aboriginal people. Deaths from land transport accidents in the Northern Territory are four times higher for Aboriginal than non-Aboriginal people.

In remote areas, primary school attendance rates have been only 64 per cent, and secondary school rates only 56 per cent. In 2003, even in Alice Springs, only 44 per cent of year five Aboriginal children achieved benchmark standards for numeracy, while 96 per cent of non-Aboriginal children did so. There are high numbers of assaults, and significantly more women than men are victims. About three-quarters of Aboriginal homicides involve alcohol in both victim and offender, and half of all Aboriginal adult admissions to Alice Springs Hospital are associated with alcohol.

In the smaller settlements, the average number of people per house in 2003 was 8.5, with 2.9 people per bedroom. In June 2004, there were 650 people on the public housing waiting list in central Australia, and the waiting time for a house was typically 4 years. The market rental rates for one- and two-bedroom units in Alice Springs in 2004 were more than the average weekly income of most Aboriginal locals.[106]

Arguing for clear thinking about the causes of these problems, Aboriginal activist and academic Marcia Langton writes: 'It seems almost axiomatic to most Australians that Aborigines should be marginalised: poor, sick, and forever on the verge of extinction. At the heart of this idea is a belief in the inevitability of our incapability – the acceptance of our "descent into hell" This is part of the cultural and political wrong-headedness that dominates thinking about the role of Aboriginal property rights and economic behaviour in the transition from settler colonialism to modernity'.[107] As part of that clearer thinking, she and Cape York leader Noel Pearson have tackled the need for Aboriginal people to be more responsible about alcohol head on, but have made it clear that the rest of society has to help with this effort.[108]

There are two counterpoints to these terrible statistics.

The first is historical. There is an almost subliminal presumption in much discourse that Aboriginal people are somehow not predisposed to engaging with a western economy. This perception was initiated when Aboriginal pastoral workers vanished unexpectedly on 'walkabout' – in fact, as we have seen, they were carrying out cultural duties that reinforce survival knowledge – and has been reinforced by stories of the impacts of welfare 'sit-down money' in recent years. It may be a little hypocritical to criticise a few weeks of cultural walkabout each year when the mainstream European/Christian-derived work system itself includes 52 Sundays and other 'holy-days' off for worship. But, in any case, this does not accord with a history that is too easily forgotten. Despite the massive disruptions, massacres, contempt for culture and forced migrations, many Aboriginal people built a close engagement with the new economy in which they found themselves engulfed.

Although the pastoral industry made plenty of money from Aboriginal labour, the relationship was often mutually respectful and beneficial. When the Western Australian Governor remarked to Gascoyne pastoralist Bill Cream around 1917 that 'with God's help you've done a wonderful job here on Bidgemia Station', Bill's riposte was 'Oh no, that's where you're wrong. God never came near the bloody place. It was me and me blackfellers did the bloody lot on our own'.[109]

Only 50 years ago in north-western Australia, there was an Aboriginal labour shortage during the mining boom mentioned in Chapter 3. Aboriginal people in the Kimberley were almost fully employed and pastoralists were obliged to invest in ever higher wages to keep them from moving to more attractive jobs in the towns, driven by the mining boom that financed both private and public investment projects. Labour shortages were so bad in the mid-1960s that farmers at Kununurra 'went to the extent of importing native labour from outside the State'.[110] Only in the late 1960s did the coincidence of a fading mining boom, equal wages legislation and changing pastoral costs of production trigger the high Aboriginal unemployment that we think of as normal today.

It was also not as if Aboriginal groups with their own financial capital did not try to develop

businesses. Around the same time in Western Australia, the Northern Mining and Development Corporation (established in the 1940s) and the Pindan Group, among others, both sought to develop mining enterprises in the Pilbara region. However, as researcher Tony Smith says, 'such enterprises were ruthlessly impeded in their operations because competing commercial operations, especially those owned by white pastoralists (who enjoyed strong links with government through agencies including the Pastoralists and Graziers Association), were greatly assisted by government policy of the time. Policy was aimed primarily at preventing the development of Aboriginal businesses as a competitive threat'.[111] How different the region might look today had Aboriginal mining enterprises developed at a time when there was a much smaller skills gap between local people and industry needs – skills that might have kept pace in the intervening period.

The second counterpoint is contemporary. Among the gloom there are many success stories in remote Australia, often involving new alliances between Aboriginal and non-Aboriginal concerns. We will meet some of these through the pages of this book – such as the successful tourism enterprise at Titjikala, south of Alice Springs; the Indigenous Pastoral Program (IPP) in the Northern Territory (formed by agreement between the Indigenous Land Corporation, the Central and Northern Land Councils, the Northern Territory Cattlemen's Association, and the Northern Territory and Commonwealth Governments); and the increasing numbers of Indigenous Protected Area agreements carrying out effective conservation works while providing Aboriginal employment in remote areas. Indeed, recent findings show that Aboriginal health is actually better in very remote settlements than in larger centres[112] – probably due to a better social environment, better family support, a healthier diet, more exercise and lower rates of substance abuse.

Nicolas Rothwell expresses the many levels of ambiguity in all this in his article 'Parallel stories'.[113] He recalls visiting Kintore, a small settlement in the Western Desert established in the late 1970s. Kintore was an early centre of the art movement that ran into problems with petrol sniffing and grog running in the 1990s. But he describes a quiet revolution of joint black and white resolve in the early 2000s, supported by better policing and the community's sense of purpose in activities such as collecting funds for their own dialysis unit. Now, he says, the community is quiet at night and the school is well run. However, there remains an uncertain sense of 'a threatened enclave holding out against the smothering pressure of the outside world'. In truth, the good and the bad coexist in Aboriginal communities, as in all communities.

The problem is that the success stories are all too often undermined by factors related to the desert drivers. They may be dependent on individuals, and collapse when these people leave the small populations. They may require distant investor interest, which waxes and wanes beyond local control. They may even be overpowered by poorly directed goodwill – when distant bureaucracies seize on a good thing and over-promote it. Understanding why this occurs is a key theme of this book. Indeed, we will argue that many problems in today's Aboriginal communities are fundamentally more related to remoteness than to Aboriginality, although cross-cultural chasms certainly exacerbate the misunderstandings and disinterest in Aboriginal affairs, and alcohol has fuelled their impact.

One misunderstanding is the idea that Aboriginal culture is somehow singular and static. In reality, it is diverse – every language group had its own special features – and evolving. In fact, there is good evidence that Aboriginal cultures were evolving before Europeans arrived, as one would expect. This was a result of both external influences and internal changes. External contacts brought in Macassan artefacts, technology and language from the north. Internal cultural evolution included some central Australian tribes taking on the use of the skin name system, which had only been adopted in the western desert a few decades before anthropologists started talking to these people in the late 1800s.

Such evolution continues today, and will undoubtedly continue to do so in the future. As Luritja leader and Northern Territory politician

Alison Anderson said (in discussing Aboriginal and Australian views on the rights of minors), Aboriginal cultural law 'is part of a living culture, and like all living cultures, Aboriginal culture has the capacity to adapt and evolve in response to change'.[114] Elsewhere she has commented that 'It is the principles underlying our governance which remain fixed'.[115]

She also points out that Aboriginal Customary Law has sought to be flexible enough to accommodate western law, far more than the reverse. Yet, by solving their own challenges of survival in desert conditions on the continent we now share, Aboriginal culture has much to teach us.

In the Tanami Desert, the Warlpiri recognise five major pillars to their culture – *walya* (Land), *kuruwarri* (Law), *jaru* (Language), *warlalja* (Kinship) and *juju/ manyuwana* (Ceremony).[116] Their term for the way in which these pillars are inseparably linked together as a way of life is *ngurra-kurlu*. Local elders see the interpretation of *ngurra-kurlu* as a beacon for their own culture, but also as a source of strength for governance, land management and law and order in a cross-cultural context.

Far to the west, the Martu people have a similar integrative concept of *kanyininpa*, encompassing *ngurra-Martu-jukurrpa* ('land-people-dreaming') in a highly interdependent way.[117] In central Australia, Eastern Arrernte elder, M.K. Turner, expresses her people's relationship to land in her poster *Everything Comes from the Land*: reminding her people, but also encouraging other Australians to understand it. [118]

Further south, researcher Diana James reports on an enormous effort to reach out across the cultural divide to develop 'cultural convergence' between Aboriginal and non-Aboriginal perceptions and values in the practice of tourism on Anangu Pitjantjatjara lands across the Northern Territory– South Australian border. They aim to achieve this *tjunguringanyi*, or coming together, through performance – also an important part of the Walpiri *ngura-kurlu*, and of course fundamental to teaching through story-telling in any culture. *Tjunguringanyi* 'does not mean assimilation but rather reconciliation by coming together with respect for difference'. [119]

These, and many other, Aboriginal groups are seeking to teach their philosophies to their next generation as well as to the wider Australian community. There is much to be learned. As a spin-off, enabling desert people to build on these lessons themselves offers new hope for Aboriginal culture too. As Anmatyere elder Rosie Kunoth-Monks said in Alice Springs, 'One of our problems is that everyone else is trying to think of the solutions for us instead of resourcing us to learn lessons and make mistakes on our own. What I am finding in the shared journey through the desert knowledge work is that we are able to learn side by side more'.[120]

James cites Anangu man Dickie Minyintiri: 'Like us today, we call each other blackfella and whitefella now, because we're strangers. But later, when we know each other, we'll all just be people'.[121]

4.4 Waste the wisdom?

So why should Australians worry about this ancient culture vanishing off the face of the Earth? Is that not the fate of ancient cultures everywhere? Is that not just 'progress'? Of course, there are moral and ethical reasons; there is the simple reality that a fascinating diversity of human experience is being lost, which is as valuable to some people as any biodiversity; and there are strong arguments related to supporting the self-identity of our Aboriginal citizens as they make their way in the world of the future. Any one of these reasons is strong enough to justify a great deal more effort to prevent further cultural losses – accidental or deliberate.

But here our case is different. The many Aboriginal cultures of the desert represent an extraordinary repository of diverse knowledge about how to deal with an unpredictable, variable, resource-limited world – a world not unlike the one the whole of humanity is heading into in the twenty-first century. Of course, the expression of these issues today is very different from that faced by nomadic Aboriginal

leaders a thousand years ago. As a consequence, some pieces of knowledge will only be interesting cultural artefacts. But others, suitably re-considered, have acute relevance for today – and for tomorrow. Usage knowledge about species and practices is most likely to be useful locally: helping future desert people build their enterprises and do their management. But other forms of knowledge are universally relevant.

With ozone holes, water restrictions, climate change and disappearing species, only a few people today would deny that the future of humanity is tied up with the life support systems of our environment. This would have seemed strange to the bold fathers of the industrial revolution in eighteenth century Europe, engaged as they were in bending nature to man's will. But, today, the underlying worldview of Aboriginal people – which sees humans as a part of the world, their fates entwined – is crucial for the future. The underlying holistic concept of many hunter-gatherer cultures is the same, and they all use a similar way of coming to appreciate it. This is through memorable storytelling about common aspects of their day-to-day lives, particularly as retold to children. It is all aimed at consciously creating an identity for people that includes respect and responsibility for the world's natural resources, and provides guidance for purposeful actions to meet these responsibilities in our daily lives. Such thinking could inform a new set of Australian stories and songlines.

Our whole world is becoming more unpredictable, more variable, more divided between rural and urban, more mobile and more resource limited. One does not have to travel barefoot across the desert to appreciate the value of Aboriginal ways of doing things in a variable and resource-limited environment. Our great opportunity lies in opening our minds in a genuinely appreciative, two-way dialogue about how desert peoples traditionally coped with these environments, and translating the lessons thus learned into the modern world.

Plate 1: Key desert strategies (i): To respond to uncertainty driven mainly by rainfall, many desert organisms escape in time – they are ephemeral. Desert sands bloom profusely in some years (a) with mixtures of poached egg daisies (*Myriocephalus stuartii*) and other ephemerals, including (b) these yellow daisies (*Senecio gregorii*). Brief good conditions also spur other organisms to sudden profusion, such as (c) caterpillars here stripping a bean tree (*Erythrina vespertilio*). Extreme variability caused by floods on the rivers of the vast Lake Eyre Basin, such as (d) near Goyder's Lagoon, can trigger massive but short-lived production by algae, creating extensive algal mats that then support a whole food chain. Photos: Mark Stafford Smith (a–c); Stuart Bunn (d).

Plate 2: Key desert strategies (ii): other organisms respond to climatic extremes by persisting against the odds. Large perennial plants, like (a) this majestic ghost gum (*Eucalyptus papuana*) in the James Ranges, invest in deep root systems to tap water in dry times; on sandplains, by contrast, (b) the spinifex and desert fringe myrtle (*Calytrix longiflora*) invest in extensive roots that are more shallow, but with the same aim of ensuring a water supply in dry times. Animals achieve the same by foraging over large areas, like (c) the fat-tailed antechinus (*Pseudantechinus macdonnellensis*) that forage individually over extensive rocky areas in drought, and different species of termites (d) in the Tanami desert (mound) and in the mulga lands (inset) which have developed specialised digestion processes as well as working together in vast colonies to build their storehouses of dead grass. Photos: Mark Stafford Smith.

Plate 3: Key desert strategies (iii): some organisms escape the desert by finding refuges within it that are less harsh or variable than the rest. Such refuges and their dwellers may be obvious, such as (a) these sedges at Running Waters in the Krichauff Ranges, or quite small and cryptic, such as (b) this little fern in a rock crevice. Other organisms are mobile, being nomadic within the desert like (c) the budgerigars that roam the inland, at times in vast flocks, or the many waterbirds that move between ephemeral desert lakes and the coast such as (d) this rich mixture of pelicans, stilts, herons and others on a waterhole near Birdsville in Queensland after the 2009 floods. Photos: Mark Stafford Smith (a, b); Wayne Lawler/Ecopix (c); Steve Wilson (d).

Plate 4: Key desert strategies (iv): a surprising number of desert organisms partner up with others to survive. One well-established persistent species can help many others, effectively reducing the level of variability that they have to cope with. For example, (a) River Red Gums (*Eucalyptus camaldulensis*) have their roots in a reliable water table below the sand, and support many other species, like the mistletoes (*Amyema miquelii*) in this picture. Less obviously, (b) the unfairly named Thorny Devil (*Moloch horridus*) exploits the reliable but sparse resource of ants. Other species partner to control the flow of the resources (particularly water) on which they depend, like (c) these mulga groves following the contours in central Australia. Other important patterns are caused by physical means such as (d) the gilgais, or depressions, seen from the air in the cracking black clay soils of the bed of Lake Woods in the Northern Territory. Photos: Mark Stafford Smith (a, b, d); Jo Caffery (c).

Plate 5: Desert weaknesses (i): the desert drivers mean that some strategies do not work in the desert and the others have their weaknesses. For example, succulents like cacti that need a regular replenishment of water do not survive in the Australian desert, as the occasional extreme droughts will kill them off; (a) caustic vine (*Sarcostemma australis*) is the exception that proves the rule – one of only two desert succulents, it is restricted to rocky hills where even small rains create enough runoff to keep it alive. A key weakness occurs where organisms are prevented from investing in reserves with which to survive dry times. For example, (b) chronic grazing of the long-lived perennial grasses which once inhabited the central Australian landscape prevents the plants from re-investing in their root and seed reserves in good times, so they gradually fade away. Some fire is tolerated by many Australian environments, but too much in acacia shrublands (c) means that the trees cannot regrow to maturity and seed before they are burned again. Sheep tracks (d) can cut channels through patchy landscapes that would otherwise capture and retain run-off, allowing it to flow off downslope, as in this chenopod shrubland in South Australia. Photos: Mark Stafford Smith.

Plate 6: Desert weaknesses (ii): the weaknesses in the biological strategies flow through to human enterprises that deal with variability. Windmills fail in bad times (a) in the absence of continued investment in maintenance during good times. As transport routes change and mobility is lost, settlements fade away, as illustrated by the ruins of Abminga Station (b) on the old Ghan Railway line. Attraction hotspots for tourism risk damage, such as the soft rock of Chambers Pillar (c) that attracts tourists but has suffered many graffiti over the years. And persistent regional drought can steadily undermine the profitability of a whole region of sheep stations, leading to plaintive signs like this one (d) in western New South Wales in the early 1990s! Photos: Mark Stafford Smith.

Plate 7: Desert solutions (i): the desert is full of opportunities for those who know how to take advantage of them. At the base of many of these is natural and cultural heritage. For example, Aboriginal art motifs infiltrate everywhere, telling stories, even linking language and culture as on (a) this birthday cake for a book launch in central Australia, while unique natural landscapes attract tourists into (b) remote gorges in the MacDonnell Ranges. Other solutions arise from responding to the desert drivers, such as (c) the Royal Flying Doctor Service, which has provided its mobile mantle of safety to the inland for 80 years, and new communications technologies (d) that allow travellers to trade their stocks and stay in touch with the world from the remotest places. Photos: Jo Caffery (a); Mark Stafford Smith (b, c); Ken Clarke (d).

Plate 8: Desert solutions (ii): other desert solutions work with the desert's characteristics – bountiful sunlight on which renewable energy systems can be based, such as (a) these solar concentrators helping to power Ntaria (Hermannsburg), west of Alice Springs. Mobile adult learning units (MALUs) (b) can be towed from settlement to settlement and rapidly set up to deliver courses to a mobile population. Careful use of local water supplies results in (c) this efficient horticulture in Alice Springs where industry and the local community agreed that the extraction rate should stay very low because the resource is only recharged very slowly. And humans often find desert solutions that involve flexibility, building on similar flexibility in desert organisms, like (d) this oat grass (*Enneapogon avenaceus*) that is a short lived ephemeral in poor years, but hangs on and flowers again if the wet times continue. Photos: Libby Kartzoff (a); Alicia Boyle (b); Mark Stafford Smith (c, d).

5

Living off a lean landscape

Roy Chisholm runs the 4450 square kilometre Napperby Station in central Australia, and now checks the state of his cattle water points by remote monitoring. The data include water levels, pump performance, rainfall and even photos. 'I even checked on my bores and water supplies from an internet café while holidaying in Brazil', he says, 'How's that for remote property management!'[122]

The natural resources of the desert are its greatest asset. Whether they are harvested for traditional foods and medicines, processed by microbes in the gut of cows and sheep, mined as precious substances, or observed and treasured through tourism and conservation, they are the source of the desert's competitive advantage. It is the mystique, rarity and purity conferred by remoteness, harshness and isolation that often gives desert products their peculiar qualities and desirability. Looking after them, with all the tools that both ancient knowledge and modern technology can provide, is vital to the future of the outback.

For the past century, the main land uses in Australia's deserts have been grazing and mining, with tourism and conservation a minor component. Today, there are new emphases: more integrated regional management of weeds, feral animals and fire; new approaches to conservation – some that engage Aboriginal rangers and pastoralists, and others that depend on a burgeoning non-government conservation sector; and prospective bush foods industries.

All these uses illustrate the main strategies for dealing with unpredictable resources. The pastoral industry as a whole persists by using cattle and sheep to harvest sparse fodder over large areas; individual pastoralists use the persistent, ephemeral, nomadic and exploiter strategies to do this. Mining and tourism industries focus on the resource-rich hotspots, with individual operators being refuge dwellers or exploiters, depending on their approach. The wild harvest of bush foods is ephemeral, driven by the seasons, but the move to desert horticulture will focus supply on the rare pockets of reliable water resources – another refuge strategy.

Much has been learned and much has changed in the deserts, yet much remains the same. We will explore the successful responses, but also how some of the other desert drivers upset the learning.

5.1 Grazing the land, the good, the bad and the different

Squatters, pastoralists and graziers were the pioneers of European occupancy of the Australian outback, along with prospectors and missionaries. Living in conditions of extraordinary hardship, they earned the right to be called battlers, even if some of the battling was very harsh on the Aboriginal inhabitants who had long preceded them. Fanning out from Sydney, Perth and Adelaide, and later the Queensland coast, from the early 1800s, waves of settlers expanded inland and across the north to the Kimberley until even the remotest parts of central Australia were settled and grazed by 1900. Sheep numbers in Australia rose from 20 million in 1860 to about 170 million at their peak in the 1970s, and have since been trending downwards slowly. Meanwhile, beef cattle rose from less than 5 million in 1860 to a peak of 29 million in the 1970s, but with a renewed overall trend upwards today.[123] The national trends are paralleled in the grazed arid and semi-arid areas, with cattle (and, in some areas, goats) replacing sheep as terms of trade change.

5.1.1 Wreckage and redemption

In the early years, dreadful damage was done to desert ecosystems by grazing, which led to collapses in the number of stock that could be carried and to significant human hardship. In their 'Learning from History' study, ecologist Greg McKeon and his colleagues document 'degradation episodes' across the outback (Box 11). The study shows that political, social and economic forces encouraged pastoralists to do many of the things that Chapter 3 showed you should not. They ran down the reserves of persistent perennial plants before big droughts, damaged the self-organising plant communities, wrecked the refuges of the refuge dwellers and generally reduced the vitality of the landscape below the level at which other persistent organisms could cope (Plate 4b, d). In some cases, they never allowed the plants to re-establish, so the cycle could not start again.

There was also human pain. Author Jill Ker Conway wrote of her family's experience in western New South Wales during the 1940s drought, in which her father was killed in an accident trying to improve their water supply, and her mother was eventually forced off the land. She recorded her feelings as she and her mother left their property in August 1945:[124]

'I did not understand the nature of the ecological disaster which had transformed my world, or that we ourselves had been agents as well as participants in our own catastrophe. I just knew that we had been defeated by the fury of the elements, a fury that I could not see we had earned.'

Rhonda McDonald, who lived through the 1960s drought in the Gascoyne and later wrote a history of it, recorded:[125]

'The drought dragged on for seven long years and before the devastation was over, it took its toll of man as well as beast. Looking over the history of the stations it will be noted a number of the original pioneers passed away during that time, even though many were not old men at the time.'

Pastoralists did not set aside sufficient fodder reserves to survive the extreme bad times, either through ignorance or, in some cases, because government policies discouraged them from doing so. Traditional Aboriginal people would not have survived long in Australia either if they had taken such approaches, and watching the changes must have broken some traditional elders' hearts. Brian Marks, an Arabana elder, put it this way: 'That was all big cattle station ... They ate it all out ... When you look at all the country like [that,] just make you sorry'.[126]

McKeon also documented how all the degradation episodes led, belatedly, to learning about the nature of the deserts. There were Royal Commissions, government reports, breast-beatings in parliaments – most of which was forgotten once it rained again. But individual pastoralists learned from their experiences.

These stories of learning are powerful. One of the earliest was published in the *Adelaide Advertiser* newspaper by pastoralist Peter Waite in 1896. He argued that stock should be spread out more widely by providing extra watering points, and remarked on the need to allow country to rest and recover:

'intelligent lessees find it necessary to spell country after it has been stocked for a time'

BOX 11: HOW THE DESERT DRIVERS COMBINE TO CAUSE DEGRADATION[127]

Queenslander Greg McKeon and his colleagues carried out an immense 'Learning from History' study to understand eight major land degradation episodes in different regions of inland Australia between 1880 and 1990 (see map). Each episode caused great damage to the land and hardship to animals and humans. Each became apparent in a major drought period, but was caused by the build-up of preceding biological, social and economic effects. The examples are scattered across the continent and through time (see map), but there was the following common sequence of events from 10 to 15 years before the drought to 10 to 15 years afterwards:

1. In the first 10 to 15 years, stock numbers and other herbivores (e.g. rabbits) increased in response to a period of mainly above-average rainfall years.

2. During this benign period, intermittent dry seasons resulted in short bouts of heavy grazing, inflicting long-lasting damage on perennial vegetation.

3. Next, a rapid decline in prices caused many graziers to hang on to stock in the hope of a market recovery; the extra mouths caused further loss of vegetation.

4. Then, a longer sequence of real drought years resulted in rapid decline in grass cover, revealing the extent of previous damage and accelerating degradation.

5. Government surveys, inquiries and Royal Commissions followed, documenting the economic and environmental problems, but only when it was too late to prevent them.

6. After the drought broke, vegetation recovered in some cases, usually as a result of sequences of above-average (La Niña) rainfall years and low stock numbers, so plants could rebuild reserves in seed banks and root systems. Recovery often took several decades – but in some cases never happened.

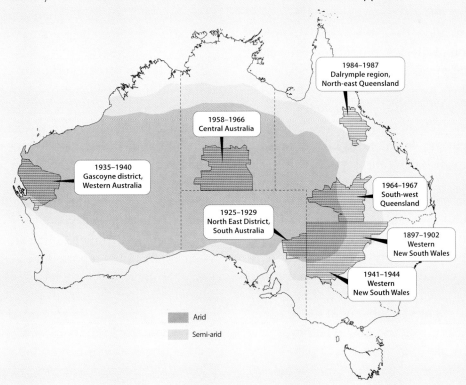

His observation echoes millennia of stored Aboriginal wisdom about the importance of allowing refuge areas to recover during good times so that they can support plants, animals and humans in times of drought.

Between 1890 and 1910, pastoralist Sidney Kidman worked out that one way of managing droughts was to move stock around the country. Because droughts could be quite widespread, this had to be at an enormous scale, so he acquired a string of properties from South Australia to the far north and up the channel country into Queensland for the purpose. In this way, Kidman was obeying another of the key laws of desert survival: nomadism. The family company still follows this strategy, now moving stock around over 120 000 square kilometres in three states and the Northern Territory.[128]

Reflecting on his experiences growing up through the 1960s drought in central Australia, pastoralist Bob Purvis wrote a remarkable account of good management principles. His style of management was to be a 'persistent local': operating at low stocking rates that could survive through almost any season without causing damage to the country or drastic declines in cattle production. He specifically sought to enable perennial grasses to build up their reserves, to allow the cattle to harvest only a small proportion of the best production over wide areas, to protect refuges, and even to reconstruct the water-flow management of the landscape by building carefully placed ponding banks that reversed the degradation wrought in the decades before he took over as manager. In all of this he had worked out the rules by which the natural system works. He says, simply, 'If you nurture the land, it will sustain you forever; abuse it and it will break you'.[129]

Other pastoralists use different desert strategies with success. In north-east Queensland Tom Mann successfully operated a more ephemeral strategy that tracked the climate in the 1990s, building cattle numbers up in good years but selling very swiftly if there was the slightest doubt about the seasons. For this, he did not wait to know whether it was dry on his own place: if there was a risk of dry times in the region, he would sell before the market

was swamped, so protecting both his profits and his country.

In western New South Wales, some pastoral blocks are only used in good seasons. These may be owned by graziers outside the arid zone, who move stock in when there is feed, then take them away again when it dries out – a true exploiter strategy.

Recent years have also seen many 'best practice' manuals emerging, mostly developed as collaborations between pastoralists, scientists and government personnel in different regions. A fine example is *Mulga, Merinos and Managers: A handbook of recommended pastoral management practices* from the North Eastern Goldfields and Kalgoorlie Land Conservation Districts, in southern WA. These outline all the ideal actions that managers in a region should take to look after their land.

Table 2 summarises many of these. Usually the manuals also go into how these issues can be reconciled with getting the best animal production, and managing your finances and property planning to make a profit. In fact, they deal with managing for all the biological weak points outlined in Chapter 3, but also show how to tackle these same weak points in the management strategy of the enterprise, such as through capitalising on the good years to build financial reserves that then help to survive the droughts.

5.1.2 Mistakes: forced and unforced

Despite all the heroes and insights, the most interesting and troubling issue is that the same mistakes continue to be repeated across Australia again and again, and often in exactly the same places. In spite of all we have learned about managing desert resources and caring for the outback, we still manage to despoil it, even in the supposedly enlightened twenty-first century.

In 2008, the Australian Collaborative Rangelands Information System reported the generally positive view that there were signs of improvement in pastoral productivity in many regions between 1992 and 2005. However, they warned of disturbing trends where stock numbers remained high despite deteriorating seasonal conditions in some regions such as the Pilbara and the Desert Uplands.[130] They

Table 2: Some key best management practices for pastoralists, and how they tackle the weak points of plant life history strategies (Chapter 3, Box 7). Numerous industry best practice manuals now list these key types of actions, suitably tailored for different grazed regions.[a]

Practices	Relation to weak points of plant strategies
Spelling (removing stock for a while) Not grazing just after drought	Allow plants to rebuild their reserves
Protect river courses and waterholes Control the distribution of stock	Preserving the refuges and better country
Conservative stocking rates Controlled feral animals (rabbits, goats, etc.)	Avoid running the system down
Cap free-flowing bores Maximise production (meat/wool) per head	Do not waste limited resources
Ponding banks and erosion control Maintain ground cover Use fire to manage grass/shrub balance	Support landscape self-organisation to retain resources
Good monitoring	Know how much resource there is

[a]See, for example, Lange et al. (1984), McCormack (1989), Morrissey and O'Connor (1989), NEGKLCD (1993), Purvis (1986), Schulke (2007); Stafford Smith and McAllister (2008) note other sources.

also observed that there was no evidence that biodiversity decline had slowed. The sustainability of Australian pastoral country remains at risk.

So, while there are unsung heroes of sustainable management of desert lands, there are also unsung villains who, for various reasons, fail to learn or apply the lessons of desert survival.

It is easy to list possible explanations – lack of information or experience, bad settlement policies, greed, complex ecology, short-sighted bank lending policies, unpredictable climate cycles, family pressures, discounting the future, silly drought policies, industry turnover, unrealistic expectations, the difficulty of monitoring change and the lack of a true Australian land ethos. All of these probably have some role to play in the continuing degradation of our desert landscapes, but equally, as we have seen, many of them can be overcome by determined and intelligent individuals who heed the desert drivers.

Unpredictable, scarce resources: there is no question that these underlying biophysical factors make management hard. In particular, the very long-term climate variability that involves big droughts and wets turning up every 20–40 years cannot be experienced more than once in a managing lifetime. This means that waiting for individuals to learn from experience is simply not effective.

Sparse and mobile population: pastoralists, their advisers and scientists are few in the desert, and there is a considerable turnover of people. This creates the benefit of having fewer people to train (less than 4000 pastoralists across the whole Australian grazed desert region, in fact), but it also creates a problem: their frequent replacements need training too, and the trainers themselves turn over just as fast. Knowledge and experience are thus frequently lost.

Remoteness: pastoralists have to work hard to get feedback from their eventual consumers – the general public – but this is changing with better communication. Today, pastoralists know about the public's environmental concerns in ways they did not 100 years ago. However, distant decision making by banks and governments still leads to short-sighted lending decisions, perverse drought subsidies (see Box 12), and constraints on multiple land use that ignore local conditions.

Local knowledge and research: even though there has been more research carried out to support the pastoral industry than any other activity in desert Australia, this has still only involved a handful of researchers scattered

BOX 12: DROUGHT POLICY – GOOD INTENTIONS, BAD OUTCOMES[131]

In 1990, the Australian Government adopted an excellent report by the National Drought Policy Review Taskforce. This laid out a clear vision of how Australia's farming and grazing industries should become more self-reliant with respect to climate variability, which has been reiterated by more recent reports.

However, under acute political pressure caused by a severe drought, in 1994, the Government introduced the idea of Exceptional Circumstances, which came to dominate drought policy. This idea was that events which occur once every 20 to 25 years, on average, may require special support. As a result, many regions in southern Australia were declared in Exceptional Circumstances continuously from 2003 until at least 2007, and some for as much as 8 years out of the previous 15. The exception again became the norm. Worse, under climate change, droughts will occur more frequently.

Subsidies during drought – and some related tax measures – are supposed to help pastoralists, but actually encourage them to damage lands leased from government in exchange for short-term financial benefits also paid by government. These undermine one of the main weak points of strategies to deal with variability (Chapter 3) – the building of reserves in good times. An exception is the 'Farm Management Deposit': this measure allows pastoralists to build up financial reserves in good years which are not taxed at that time – correctly avoiding the weak point.

The ultimate aims of drought policy are to ensure that farming families do not suffer undue hardship (an aim that is better achieved through social security) and to help non-viable farmers to leave farming with dignity before they wreck the country. There is little evidence that the latter purpose is being met by policies of recent years.

The problem is that government is distant from the problem, so that a quick fix – which buys off bad publicity in the short term – is more important in politics than bad outcomes for people and the environment in the longer run. The solution is to support good planning for climate variability, including the building of reserves in good times, and then ensure that no farming family suffers undue hardship through social security.

across three-quarters of the continent. Lack of understanding among city policy makers originally contributed to over-expectations and bad settlement policies – and the continued absence of comprehensive monitoring perpetuates these. Only recently have better ways of combining formal research with essential local knowledge come to the fore. Oddly, the early pastoral industry depended heavily on Aboriginal knowledge, yet today most of the industry is wary of its value, perhaps understandably worried about interference with their independence.

Different culture: the close-knit culture of the pastoral community in the region cuts both ways. It can encourage best practice management of desert resources – but it can also act against this. If people decide an idea should spread, it soon does, but there can also be a shared battler mentality against the outside world, and a reluctance to act against a peer who is obviously doing the wrong thing. Although individuals have a strong ethos of wanting to hand their land on to the future in better condition than they received it, they are only human; when times are tight and they want to get kids through school, they may pressure the country. The only way to avoid this practice of discounting the future is through some degree of imposed obligation to behave otherwise.

BOX 13: KNOWLEDGE ALLIANCES TO WARN OF OVERSTOCKING[132]

Pastoralists do not like being told how many stock they should carry, saying they should be free to make their own decisions, but we have seen this is not enough by itself. Peter Johnston spent a decade working on a 'safe carrying capacity' formula that combines local knowledge with regional science and policy.

He asked pastoralists to identify how many stock could be run on different types of country in the long term. Then he used models to estimate how much country fell into each category and checked their estimates of productivity against the local knowledge. Eventually, an estimate emerged for how many stock each land unit, paddock and property could safely carry, based on the best of the local knowledge and supported by science. Neighbours know the estimates, and one day they may be used to discipline those who do not stick to them. For now, enforcement is down to peer pressure – backed by community consensus on the calculations.

These same estimates have flowed up to a regional modelling system called DroughtAlert, which uses the model 'AussieGRASS' and seasonal forecasts to forewarn graziers when they are carrying too many stock. This helps to highlight where graziers or state agriculture staff should swing into action. The fusion between local knowledge on the ground, regional science and broader policy has the potential to reduce the risk of major land degradation events in future. It is also a modern example of consensus knowledge at a regional scale that could persist over generations – in much the same way that traditional Aboriginal knowledge was held communally to last through generations of uncertainty.

It is often said that it takes a lifetime of experience to really understand the pastoral country and own that understanding. But today this is not enough: the long-term climate cycles and slow responses mean that *individual* experience is quite inadequate. As Aboriginal culture showed, as we are learning again today, there is a need for codified community knowledge at a regional scale. This requires an alliance between pastoralists, government and science that passes knowledge on reliably as the individuals change. Aboriginal people did this with a complex of detailed songlines and cultural rules, and we need to develop the equivalent shared stories and ethos (with incentives and penalties) for desert land management today.

To avoid discounting of the future and deliberate greed, we need more than information and goodwill. There must be a system that imposes strong peer pressure first, and then formal sanctions when necessary. This needs to be more than industry self-regulation, and needs to be driven by regional groups that encompass and empower all knowledge-holders and resource users, including conservation, water use, tourism, local government, research and

Aboriginal knowledge. The current regional natural resource management bodies are a step towards this, perhaps though not with a consistent formal role in each state, nor with true devolution of power (see Chapter 9 and Box 20). Box 13 illustrates some smart desert knowledge that has been emerging in Queensland, where the local insights of pastoralists have been combined with scientists' models to produce a better understanding than either had alone. The principles behind this DroughtAlert system, coupled with appropriate regional governance, could control bad management more proactively in future.

Finally, the issue of distant decision making is hardest to deal with, because it is at the heart of the desert drivers. Here, the best that can be said at this stage is that it would be good if higher tiers of government helped local people to make better balanced decisions about their resources and future, but stayed out of meddling with the details. Later chapters will explore in more detail how this might happen, but the importance of local and regional involvement in caring for our desert lands cannot be over-emphasised. Desert Australia is diverse – some regions will remain dominated by grazing while others are rapidly changing as conservation, mining and Aboriginal values come to the fore. Decisions about these are best made by people who know and respect their country.

5.2 Desert conservation

Traditional Australian land-use practice has been to grab the accessible, watered, productive land for agriculture, and leave the remainder fenced off as a few parks and reserves, or Aboriginal land. At its best, the result is a sea of thriving agriculture, with natural landscapes confined mainly to upland 'atolls' where they occasionally survive on a large enough scale to be conserved. At its worst, the results are the eroded, acidified or salinised parts of the Murray–Darling Basin and the wheat belt of Western Australia, with tiny reserves of dwindling native species infested with weeds and feral animals. People now have to work hard to reverse this outcome.

Conservation management aims to maintain all the processes that generate biodiversity in perpetuity, but in practice it mainly focuses on two things: fixed places and particular species. It aims to incorporate and manage all the different vegetation types in their current locations, and to protect rich pockets of habitat that are known (or believed) to support rare or locally diverse biota. As a result, old-style Australian conservation management breaks several of the rules of desert survival, particularly the need for species to move across the landscape in pursuit of scarce and variable resources.

The implications of climate change in this endeavour are grave.[133] Weeds and short-lived, mobile native plants will respond to a change in climate much faster than long-lived trees, so vegetation types are not going to move with climate in any coherent way. Some species will die where they stand. In 30 to 50 years, half our current vegetation communities will no longer be recognisable in the places where they are today – or anywhere else. At the same time, the habitat for many endangered species will have moved (or disintegrated), and biodiverse hotspots will no longer support the same species that they do today. Our grandchildren will probably never see or experience the Australia we now know, though fragments may remain like living museums.

Australia needs new approaches to conservation, and these are highlighted in the desert.

The ways in which the landscapes of desert Australia differ from those of agricultural regions provide important pointers. Firstly, most desert landscapes are hugely extensive and flat: this means that small changes in climate may force organisms to move a long way. It is immediately clear that isolated parks are a poor way to plan for the future in such a vast landscape. This is true even without climate change, because we know that from time to time any given location will experience an extreme drought during which some species (persistent ones especially) may go locally extinct. If these species cannot move back in from elsewhere then they are lost for good.

Secondly, grazing is a modifying, rather than a transforming, force on most of the environment – that is, much of the grazed landscape between parks is still semi-natural compared with ploughed croplands, where native vegetation has been largely eliminated. This means that native plants and animals can live there too, or at least pass through it from time to time. Land managers in the desert regions (and elsewhere) are therefore a central part of the conservation solution – the task cannot just be left to a handful of under-resourced park rangers.

Dealing with all this requires two shifts in perspective. Firstly, we need to think about managing conservation at a regional scale, not on individual blocks of land. And, secondly, we have to stop thinking of one land manager – the pastoralist – as necessarily being a pillager and the other – the ranger – as the conservationist. These roles have to be brought together.

A project called 'BioGraze' deliberately investigated how different types of protection can be integrated across a region to deliver regional conservation.[134] At a regional scale, Biograze looked at linking up parts of grazing properties that happened to be a long way from a water point and therefore lightly grazed. The central idea in this was to pay the land manager some sort of stewardship salary to look after these areas and avoid grazing them heavily, so the whole landscape could remain connected. In this way, the formal parks and reserves would no longer be isolated atolls in an agricultural 'sea'; instead, they become outcrops embedded within a multi-purpose, but hospitable, landscape in which native species can both persist and move. Land managers then need to join forces to control other attacks on the weak points of the strategies of different species. In particular, they need to maintain a natural fire regime across the region, and to control feral animals and weeds; these are also tasks that deliver a public good for which the managers should be recompensed.

This approach can help both pastoralist and landscape. For example, we spoke to a pastoralist on a remote Nullarbor property in Western Australia who was struggling with the economics of sheep grazing. But, adjoining his lease was a conservation reserve to which a ranger had to drive out 600 kilometres from Esperance every week to check. He offered to take on the ranger's role; this would save the government money and give him extra income, which would ease the need to put pressure on the rest of his land. He was also exploring a small tourism venture that could take advantage of the improved image, thereby again reducing the economic pressure on the land. Better still, these other options were more or less independent of the climate variability that upset the cash flows from sheep production. He was trying to create more stability and security in his own income, at the same time as better protecting the overall landscape and its species.[135]

There is a challenge in all this: there is no point in the public purse paying stewardship salaries to bolster bad management. Such schemes must obtain a genuine management service for society in exchange for the payments. They cannot become a subsidy to the pastoralist to stay on the land. In fact, many operators may have to stop seeing themselves primarily as pastoralists and start seeing themselves as land managers with professional interests in both grazing and conservation.

This way of thinking creates a new role for the landholder, ranging from conventional farmer, through mixed land manager to park ranger, with the balance of roles depending on the region. In a highly productive region, most managers will focus on production from their land, but still carry out wider conservation management with regional implications, such as reducing catchment erosion, maintaining water quality, managing fire and controlling feral animals. On large leasehold properties in marginal lands, managers will have a more mixed role, supporting regional conservation values in concert with some residual pastoral production. In regions that really can not support grazing at all, managers become conservation rangers who are focused primarily on conservation, but probably balance this with some economic production, perhaps in the form of tourism. This all demands a far wider view of the role of 'desert land manager': actively integrating all these roles to achieve the appropriate balance of conservation and production outcomes, depending on the circumstances.

Of course, this is not just a model for pastoral lands. A quarter of desert Australia is managed by Aboriginal peoples, and they have native title interests over much more. In the past decade, there has been an explosion of interest in Indigenous Protected Areas (IPAs), coupled with Indigenous Ranger programs (Figure 8). IPAs are areas of Aboriginal land that the traditional owners agree to declare as having conservation value to the nation. Funding is then made available to plan the management of these lands, both in the public interest and in terms of maintaining cultural values in that land. Twenty-five IPAs, covering more than 200 000 square kilometres, had been declared as of 2008 – from the first in the desert at Nantawarrina in South Australia to the most recent on Cape York.[136] One of the largest is the Ngaanyatjarra Lands IPA – 98 000 square kilometres, declared in 2002 – which is managed for fire, waterholes, cultural knowledge and, in particular, the threatened Black-footed rock wallaby whose numbers are now increasing. There needs to be long-term funding for Indigenous Rangers who carry out the management required – creating real

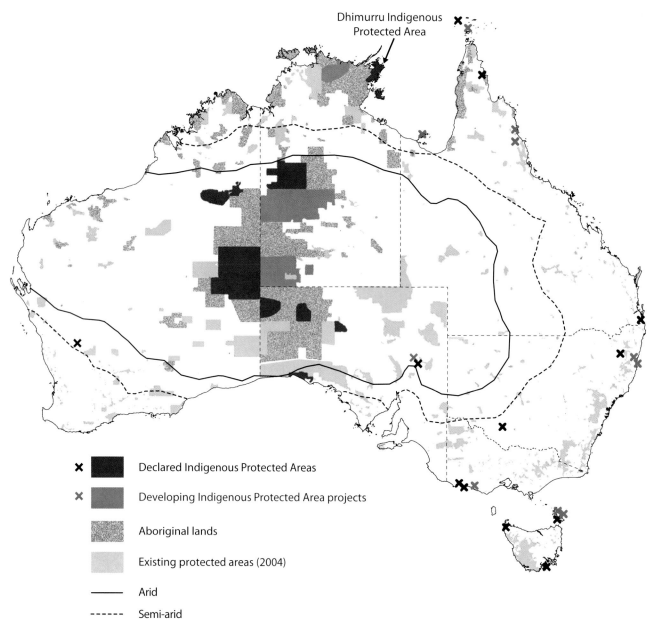

Figure 8: Locations of Indigenous Protected Areas in 2008, in the context of other protected areas and other Aboriginal lands (x marks sites too small to see at this scale).[137] In the Top End, the Dhimurru Indigenous Protected Area, with its ranger program, has been estimated to cost about $1.40/ha, compared with the cost of managing Kakadu National Park at $8.90/ha. In western New South Wales, stewardship salaries paid to pastoralists to carry out conservation management in a trial program were estimated to cost $2–4/ha compared with typical costs of managing parks in the region of about $20/ha.[138] This is a powerful point – providing that the required conservation outcomes can be achieved, it makes a lot of sense (and may be cheaper) to pay people who want to live in the area to do the management than to have to bring in specialised staff from outside.

jobs in these remote areas – with a potential career pathway into regional conservation management.

Governments are still confused about what stewardship means in relation to managing natural and cultural heritage values. Indigenous Rangers were initially paid from an untargeted employment program. Many of these funds have now been transferred into long-term programs for conservation that can fund management jobs properly. This helps to clarify the intention to pay Aboriginal people to look after land and culture on behalf of society at large. A similar confusion over stewardship remains in the pastoral industry; there is an urgent need to sort it out in order to develop the right policies. The signals and incentives to pastoral managers to look after the landscape on behalf of society have to be clear and unambiguous. This is starting to appear in programs that pay pastoralists in western New South Wales, and Aboriginal rangers in the Top End, to look after their landscapes.

There is a further role for Aboriginal people in contributing to conservation of cultural values across desert lands in general. This depends on pastoral managers recognising the value that may be added to their management by involving traditional owners in decision making where it relates to cultural values. Such suggestions naturally arouse suspicions among pastoralists about meddling in their management. However, there could be great benefits for all in having a regional forum at which all land managers contribute their ideas and aspirations for how the region is cared for and managed, and how common problems such as fire, feral animals or weeds are dealt with. In the long run, such forums, growing out of the natural resource management groups, could give desert people the degree of autonomy from remote central government that they need in order to run their own affairs and care for the land.

In summary, a new model for conservation is emerging in desert Australia, involving paying pastoralists (whether Aboriginal or not) to look after grazed landscapes on the one hand, and paying Indigenous Rangers to care for ungrazed Aboriginal lands on the other. These can be integrated at a regional scale with the growing number of private conservation reserves, and with leases owned by sympathetic mining companies, to lead the way in developing a holistic approach to conservation management for these lands that make up the greater part of Australia. All this requires a big shift in culture and thinking, but with enough goodwill, persistence and a common interest, people can be brought together, as the next story about fire shows.

5.3 Burning issues

The Tanami desert stretches for a thousand arid kilometres of sand dunes and spinifex to the northwest of Alice Springs. Botanist Peter Latz believes that only a few thousand years ago this was a shrubland, with patches of tall eucalypt woodland on old river drainage channels, which were mostly dry but with some water still flowing beneath the surface along the old river routes. In his book, *The Flaming Desert*, Latz argues that fire has helped the flammable spinifex to invade this vast land, and that spinifex has helped fire to dominate the landscape.[139] Fires race through the spinifex on a hot day and burn off the edge of the woodlands, killing the sensitive mulga (Plate 5c). Then the spinifex invades a bit more and burns again before the shrubs can recover and seed. So, the alliance of spinifex and fire inexorably erodes the remaining patches of woodland.

A mixture of a drying climate and human burning probably initiated the cycle – and only continued management of fire can now mitigate its most intense effects. When Europeans arrived in the area, Aboriginal people were still burning in patches, for hunting, signalling and just 'cleaning up the country': fire-stick farming to ensure future food resources and to control the size of fires when they were lit. But newcomer pastoralists (settling on wistfully named properties such as 'Newhaven') were concerned not to burn off the fodder that their cattle depended upon, so they sought to suppress fires.

Using satellite imagery, fire scientist Grant Allan calculated that the area of central Australia that burned in 2000–2002 was bigger than Italy and one and a half times the size of Victoria![140] Most fires were near roads and tracks, and were started by people: the result was conflict between pastoralists, Aboriginal people, government fire-management agencies and conservationists over how fire should be managed.

As anthropologists Petronella Vaarzon-Morel and Kasia Gabrys found, today, Warlpiri people in the Tanami Desert are concerned to maintain fire practices, because it is part of looking after country and keeping it productive in a culturally appropriate way.[141] In fact, as the two researchers say,

'Fire is ever present in the lives of Warlpiri. It is frequently used for cooking food, for warmth and light, to clean campsites and in the production of artefacts such as spears, shields, coolamons (carrying dishes), necklaces and wooden sculptures. People also use fire for safety reasons, such as warning off snakes, dingoes and malevolent beings. Fire is of symbolic significance in rituals which mark transformations at different stages of the life cycle such as birth, initiation and death, and in Jardiwanpa, a fire ceremony performed to resolve social conflict… Fire is also utilized in hunting and gathering activities and in other ways that affect the physical environment.'

There are several songlines about fire – Fire Dreamings – in the southern Tanami, which are still of great significance to the local Warlpiri. However, some Warlpiri who were active in the pastoral industry when they were younger are now reluctant to burn even on Aboriginal land in case it might damage property or escape and burn fodder on surrounding pastoral stations. As a consequence of this type of concern, active experience with fire is diminishing in the next generation. Knowledge and confidence about using fire is being lost. At the same time, people can see the benefits of linking traditional cultural management to today's fire-management practices to help them access their country, to look after it and to train young people.

There is a similar diversity of views on the surrounding pastoral stations. Kirsten Maclean interviewed several pastoralists[142] and found that while some pastoralists prefer to exclude fire from their leases because it threatens pastoral grasses and infrastructure, others burn when the opportunity arises in order to prevent uncontrolled wildfires. Still other pastoralists actively burn for pasture management purposes whenever the weather allows, to promote the diversity of native grass species. Consequently, some pastoralists are in conflict with Aboriginal people who start fires on their leases; they may even be in conflict with their pastoral neighbours who use fire. On the other hand, she cites one pastoralist explaining that Aboriginal use of fire on his lease can be mutually beneficial because:

'I don't have to [burn] a lot myself, me [Aboriginal] mates here do it for me! … I think that is too why we haven't had a lot of big fires because of all their hunting tracks… If they started lighting up in our good grasses it would be [a problem] but while they are just burning on the Spinifex [it] doesn't matter.'

In the same region, the Northern Territory's Bushfires Council is also active, and there are now both private and government-run conservation parks. Everyone has different views on fire. In fact, many people are confused about the seemingly contradictory positions relating to use of fire held by scientists and pastoralists, and enshrined in legislation, all of which have been changing over time. For example, Vaarzon-Morel points out that there is Northern Territory legislation that prohibits burning at certain times of the year; yet other legislation supports Aboriginal customary practices such as burning on Aboriginal Freehold Land at any time of the year. She even found that most informants, whether Aboriginal or non-Aboriginal, had no idea which legislation had the greater power to be enforced and where.

All in all, there is a wonderful opportunity to bring together these different viewpoints to mutual benefit. The 'Desert Fire' project[143] has been seeking to do just this. Its goal has been to develop a collaborative fire-management strategy that weaves Aboriginal fire knowledge and practices in with those of non-Aboriginal interests in the Tanami region. It recognises that such a strategy needs to bring a diversity of views and experience and knowledge to the table, with genuine two-way partnerships and learning, and creative and flexible approaches to develop outcomes of mutual benefit. And these benefits are many: controlling fire away from critical pastoral and conservation lands; meeting habitat needs for more diverse and appropriate fire regimes; maintaining traditional culture; and reducing costs for government in fighting unplanned and potentially dangerous wildfires and in getting better conservation results. This is a real win–win–win win possibility!

However, fires only occur once or twice in a decade in the desert, unlike the annual fires of the Top End, so these regional partnerships need to sustain the knowledge over long time periods to avoid a loss of interest and preparedness between the big fire years. Given that climate change is likely to create the conditions for more fires in intermittent future years elsewhere in Australia, and indeed around the world, learning how to resolve the conflicts that fire inevitably causes is vital far outside the desert.

5.4 Tucker, tourists and treatments

New industries based on renewable natural resources are germinating in the desert. Tourism has been around for a generation or so now, and depends on keeping heritage values in good condition. Bush foods and bush medicine industries are being developed. And the desert has long been a source of natural products such as sandalwood. It has taken us a hundred years to start to get the principles for sustainable grazing embedded in pastoral management, but there is no excuse now for making similar mistakes in newer industries.

The tiny community at Titjikala, located 120 kilometres south of Alice Springs, has been a leader in several developments. Yankunytjatjara, Luritja and Arrernte people live together at Titjikala. Over the past decade, they have begun to record and save their cultural knowledge, and then build on it in developing a tourism enterprise with an outside company ('Gunya') and projects on bush foods and medicines with university researchers (the 'Plants for People' project).

Titjikala people have partnered with computer specialists in Darwin to create an online database of their cultural knowledge. Some of this knowledge is secret – or can only be revealed to people in particular families, language groups or of a particular gender – whereas other information is freely available. Today's sophisticated log-in procedures (just

think of internet banking) can provide this sort of access control. Just as the Tanami Warlpiri saw the benefits of using satellite technology in their fire management, so too the Titjikala people have found solutions in computer technology.

Thus, high tech and traditional knowledge systems can help each other. And once knowledge is recorded, many other opportunities emerge.

In 2002, Louis Evans visited Titjikala from her base at Curtin University in Perth; her nephew Harry was working there and he introduced her to Mr Johnnie Briscoe, one of the community elders.[144] Among other things, Evans is an expert in chemical analyses and Mr Briscoe involved her in discussions about gaining benefits for the community from their medicinal knowledge of plants. There has been a long history in Australia of big companies coming in and testing many species for useful chemicals: essentially using local knowledge without giving anything back. The elders and Evans wanted to establish a better model, in which the benefits would stay with the community. Over the years the resulting Plants for People project has trained several local people, recorded large quantities of local knowledge about local plants, explored the properties of some of them, worked through some complex exemplary legal issues and has been testing some products that could be commercialised. Most importantly, the whole project has involved building trust across the cultural divide and then pooling local and research knowledge to find useful products.

At the same time, traditional law hybridised with modern legal concepts of intellectual property, resulting in new approaches.[145] To avoid getting tied up in endless legalities, the Titjikala community proposed the ideal benefit-sharing principle, as described by the vivid Yankunytjatjara phrase 'ngapartji-ngapartji'. This is an expression of reciprocity and working together in a 'give and take' partnership of equals, and is now also the title of a cultural exchange and community development project based out of Alice Springs.[146] For the Titjikala project, ngapartji-ngapartji was legally interpreted (because the lawyers baulked at putting the Yankunytjatjara term in the agreements!) as the right to start negotiations in any commercial deals from a position of expecting equal shares in benefits, however those might be counted; give and take, equal partners. This is another sound approach taken from survival principles evolved in deserts.

There is a larger issue in developing new desert product industries. The risk is that, as soon as an enterprise starts to be successful, someone with outside capital buys it up, plants vast quantities of the product on irrigation and seizes the market. In doing so, they cause the price of the product to collapse and manual harvesting in remote regions to become uneconomic. Desert people then lose all the benefits, which are captured instead by city shareholders and banks. Today, researchers are trying to develop models where this does not happen. This may be by keeping intellectual property rights for good cultivars in the hands of desert people, or ensuring that desert interests have a formal share in the developing industry through partnerships, or devising ethical marketing labels that require some contribution to return to the desert. Such approaches may also apply to other industries.

Titjikala is also the site of a remarkable tourist development that is based on private investment capital. Thanks to multiple visionaries in city business, settlement management and the community itself, Gunya Tourism and Titjikala have partnered since 2004 to establish a small, high-end tourism experience, with just five luxury safari tents and the promise of a very exclusive experience.[147] The venture has a joint management board and guarantees that all jobs at the resort go to people in the community. It aims to allow the community to express their own business model and interpret that to the mainstream as needed, rather than having the mainstream model foisted on the desert. It only expects 25 per cent occupancy and was forecast to take 5 years to make a profit; in fact, it broke even after 3 years. More importantly, it has provided education and employment, and a changing sense of future opportunity – in 2003, only one student from Titjikala attended high school whereas in 2007 there were 24, who were largely motivated by realising that they needed an education to be able to work in the tourism enterprise. As Gunya says, tourism is an ideal vehicle for development because 'it builds on the natural attributes of Aboriginal people, the world's oldest story tellers'.

All this is happening in just one desert community, Titjikala, but there are examples of such constructive cultural collisions all over the continent, little recognised and inconsistently supported. Together, they emphasise that, in the desert, no one size fits all – each region has its own characteristics, and solutions must be true to these.

5.5 The desert natural resource manager of the future

The desert is diverse: from mining booms in the Pilbara, to core grazing areas such as the Barkly Tablelands, tourism-dominated regions such as central Australia, and impoverished Aboriginal homelands in the central deserts. Geographer John Holmes has formalised these differences (see Figure 9) by identifying seven types of regions and classifying 29 remote regions of Australia into these categories (five he could not place) based on their trajectories in 1994. He reviewed this a decade or so later and found the analysis had stood the test of time well. Thus we can expect the future to take different paths in these different types of regions.

In some 'core pastoral' regions, where grazing is profitable most of the time, there is little competition from other land uses, and it is likely that grazing will still be the main occupation in 50 years, even with climate change. This is particularly true of many northern cattle regions.

Other regions are marginal or are dominated by other activities and other futures, such as mining or tourism and recreation, or Aboriginal homelands. Such regions are the real desert country, where pastoral economics are increasingly marginal and markets are geographically distant. Here, past approaches no longer work; the future is all about the desert trait of having eggs in many baskets – and changes are already happening. The economics of production mean that managers have to look around for multiple sources of income, maximise the value

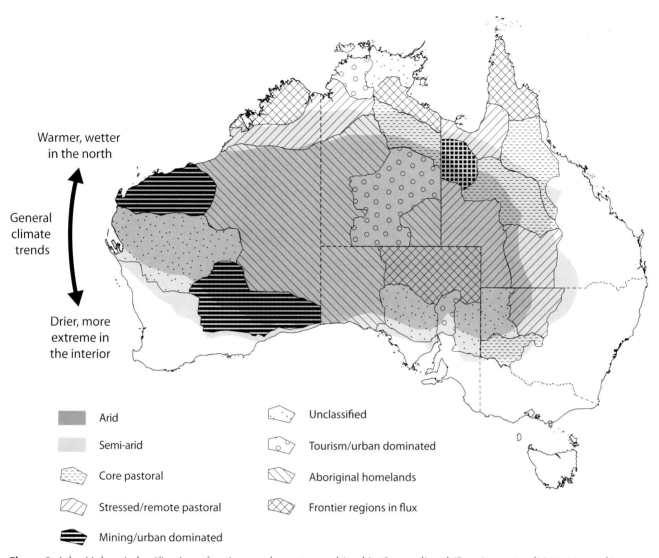

Warmer, wetter
in the north

General
climate
trends

Drier, more
extreme in
the interior

	Arid			Unclassified
	Semi-arid			Tourism/urban dominated
	Core pastoral			Aboriginal homelands
	Stressed/remote pastoral			Frontier regions in flux
	Mining/urban dominated			

Figure 9: John Holmes' classification of regions, redrawn to combine his 'Stressed' and 'Frontier pastoral' into 'stressed/remote pastoral', and show the general climate trends expected for the northern and southern parts of desert Australia.[148]

of their product and minimise their costs, whether they are managing grazing or tourism or bush foods. How will they do this?

Firstly, future resource managers are already diversifying. On Napperby Station in central Australia, Janet Chisholm has been trading bush foods since 1993 as a sideline to various other pastoral enterprises, including cattle grazing and the local roadhouse.[149] Napperby is a classic mixed property, where the different elements of the business support each other. Other properties, if given the chance through enlightened government policy, will move down the line of mixed conservation and production. Still others are engaging with the mining industry, or tourism.

Secondly, managers are making sure that what they do produce is worth as much as possible. The desert property will always be a long way from market, so transport costs are critical, whether you are shipping cattle and bush foods out or tourists in. One competitive advantage of the desert is low-cost production: you do not have to plant fodder, you can avoid many of the diseases of closely settled areas and the views are (more or less) free. To stay ahead of the cost–price squeeze, managers will use the naturalness of the desert landscape as a marketing edge in a world where everywhere else is getting less natural and half the human race is confined within vast concrete termite mounds. They will also use new technologies to stay closely attuned to what the global market wants. Just as Aboriginal artists sell their canvasses over the web from the Tanami to New York, so the OBE Beef Company at Birdsville is selling its organic beef online into Tokyo, Taiwan and New York (with Chinese, Japanese and Korean translations) from the outback.[150]

Thirdly, managers will deploy new technologies to minimise their costs. Back on Napperby Station in central Australia, Janet Chisholm's husband Roy uses remote monitoring technology to save costs on bore runs, as noted at the start of the Chapter; he is also experimenting with remote monitoring of cattle. Clever uses of radio and new mobile phone technologies allow pastoralists like him to check whether their water points are working or not when they are at home, in Alice Springs or even overseas. From this, it is a short step to controlling the release of special nutrients into the troughs to look after the animals. All this means they check bores more often but visit them less frequently: saving time, fuel and wear and tear. Pastoralists already access web-based satellite data telling them where fires are burning on their vast outback properties. Similar technology will soon allow them to see the state of their pasture growth, summarised over large areas, so they can plan stocking levels for different parts of the property, and adjust numbers accordingly. Capital-intensive fences may be replaced by fenceless paddocks, where animals have tags that direct them to stay within a particular area, providing a huge potential for fine-tuning the management of the country. Once all the animals are linked electronically, their health and weights can be monitored constantly, ensuring they produce more calves and identifying animals that are ready to go to market. Researchers have already trialled automatic weighing machines at the water points that can report animal condition remotely[151] – another part of their holistic approach to management (see Chapter 8 for other approaches to managing water).

All this smart thinking contributes towards 'precision pastoralism' – carrying out grazing with

much greater control and efficiency. The future of the world will see a rising demand for staple food crops, which will push more of the meat production into natural grazing lands, and a drying climate that will mean an expansion of these regions. The use of more sustainable and productive grazing technologies will be a global necessity, and precision pastoralism will be a significant export opportunity. Similar trends can be seen in tourism, conservation management, mining and other natural resource-based enterprises.

5.6 Lessons from the desert

Over tens of thousands of years of trial and error and learning, Aboriginal desert dwellers had established a balance with their environment. They used the scarce resources but protected them; they handled the extreme swings of climate, and they worked out how to carry the knowledge that might only be needed once every few centuries in communally held rules and stories that survive the occasional forgetful individual. In the last century or so, pastoralists too have learned from their forebears and developed their own knowledge for looking after the land. In the twenty-first century there is a need for the regional knowledge systems that could substitute for Aboriginal songlines in the modern world, and these are slowly appearing. Yet, despite the successes, there are also continuing failures, which a better understanding of desert systems can help us to avoid in the future.

As the rest of Australia, and indeed the world, faces some of the desert drivers – greater climatic uncertainty, limited water and nutrients and an increasing divide between urban and rural areas – lessons can be learned from the desert experience. In summary:

- There are many ways of harvesting scarce, variable and patchy resources, and local people need to be allowed the flexibility to develop the best approaches – just as plants, animals and nomads have in the desert.

- Growing uncertainty over the variability of climate and resource availability requires social learning systems that transcend individuals and persist across generations. Aboriginal people solved this with their systems of cultural songlines. A modern version of this means thoughtful regional alliances among industry, research, government and community interests.

- Remoteness from markets (especially in times of increasing transport costs) means focusing on greater efficiency and smart use of technology, and giving priority to high product value. The competitive advantages of remote and rural areas are largely to be found in their natural and cultural resources (Plate 7a, b), and these therefore need to be treasured.

- Remoteness from centres of political power ('*distant voice*') means that good local response strategies can easily be wrecked by the external imposition of perverse incentives and bad policies. Desert dwellers need to recognise this, and argue against these things with one voice, while sympathetic distant governments should work to minimise their accidental impacts. Permitting effective regional governance with local sensitivity is one way of doing this (see Chapter 9).

There are many ways to live successfully and sustainably off lean landscapes, as the examples in this chapter show. The world of the late twenty-first century will itself be lean, with far fewer resources to share among many more people, each demanding more. To avoid wars, famines, economic depressions and the other consequences of resource scarcity, it will be necessary to apply many of these principles to our global and personal lives.

6

Dry but smart: tomorrow's desert business

Arrernte people in central Australia have been networking the land for millennia. They traded specialist ochres for painting 1500 kilometres across the Tanami Desert from the Kimberley, and pituri for chewing 800 kilometres through the Simpson's sand dunes from western Queensland. In fact the whole continent was a vast web of dreaming tracks following lines of social relationships and trade, with goods moved immense distances on foot. Trading included axe heads, spears, shells, grinding stones, shields, boomerangs, stories and culture. It also extended to 'international' trade, with goods being exchanged with the Macassans of Indonesia and across the Torres Strait.

What has worked for desert dwellers for thousands of years may hold some valuable messages for smart desert business in the modern era. Isolated by vast distances from customers, suppliers, service providers, researchers, trainers, government offices, markets and ideas, the enterprises of the inland are latching on to new communications technology and business networks to build nationwide and global reach. In so doing, they are pioneering models for the globalised small business of the future.

In a place where the typical enterprise has fewer than 20 staff, the nearest supplier may be a thousand kilometres away and customers are spread all over a rugged landscape, or even the world, the challenges of building a viable business enterprise are greatly magnified. Yet there are over 38 000 small businesses in desert Australia: twice as many per person as the overall Australian average.[152]

In keeping with the immensity of the landscape, outback enterprises have often been on the visionary scale, from Aboriginal trade across the continent, to the cattle and sheep empires that characterised the era of European settlement of Australia, to today's world-class mining operations and distinctive eco-tourism ventures. The new vision that the outback is bringing to business practice deals in particular with how businesses communicate and link with one another. This they must do in order to muster the skills, ideas and critical mass necessary to make an impact in continental or global markets.

Reflecting their environment, a defining feature of most outback businesses is their ability to stand alone yet work together. Not only must they be able to cope with the intrinsic variability of landscape and climate and the ups-and-downs of the business cycle, but they must also be self-sufficient. When a city business runs low on production inputs or equipment, it can obtain fresh supplies close by; if it loses one customer, it may readily find another. By comparison, the outback business must sustain itself in all things, at the same time building durable bonds with customers, suppliers and business partners who may be far away and seldom met face to face. In such dealings, trust and relationships are of even greater importance than in urban business dealings.

6.1 Desert business strategies

Small businesses everywhere must deal with uncertainty in their operating environments. But, like desert plants, remote desert businesses are exposed to sources of unpredictability that are far greater than those of their city counterparts. Pastoral businesses are obviously affected by climate variability: wet years and dry. But even businesses that are not dependent on the natural environment are variable. They are remote from most suppliers and are affected by problems with transport, whether caused by floods or strikes or being last in line for limited supplies. They experience problems with finding, training and retaining staff from the small population pool. They have to make do without easy access to support services that are normal in the city, such as financial advisers, recruitment agencies or government business advisers. Their peers (and competitors) are a long way away, so their industry bodies may not be much help. They have a limited choice of accountants or mechanics or repairmen. Like desert rains, new business opportunities and investment sources, such as a mine calling for tenders, turn up irregularly and unpredictably. The customer base is narrow and it only takes a few clients to be away on holiday to close you down. Communication is a perpetual challenge.

However, as with plants and animals, many outback small businesses have learned to exploit what seem at first sight like major disadvantages; and, as for desert plants and animals, different strategies suit different purposes at different times. For some businesses, opportunities are sparse and scattered, but are always there if you are on the lookout: this suits a 'local persistent' sort of strategy. Other business opportunities appear when a new mine is opened, but are then quiet for decades: in

this sort of business you have to be ephemeral or nomadic, or build alliances to have a better control over your opportunities. Other types of businesses can sit tight, like refuge dwellers, and concentrate on small pockets of rich opportunity, such as a local tourist hotspot.

In fact, just as for the plants and animals in Chapter 3, we can see:

- *The persistent strategy* – pastoralists, hairdressers and small businesses doing the mechanics, plumbing and joinery that the big companies cannot be bothered with. There will always be low, but reasonably steady, demand for these suppliers who, if they have any sense, build up reserves to tide them over the inevitable downturn.

- *The refuge strategy* – these are companies occupying a small, but reliable, niche based on some fixed opportunity, such as tourism around Uluru or horticulture on the Ti Tree groundwater basin. These companies have to do everything they can to protect the sustainability of their resource.

- *The dependent strategy* – businesses whose reliable niche is provided by bigger players needs, such as mining companies, government defence contracts or outsourcing of computers and social security. These niches can be disrupted by mineral boom cycles or changes in government policy. These businesses must use their niche as a springboard to larger markets and eventual independence if they want to survive outside the umbrella of the big player.

- *The ephemeral strategy* – this is less common among businesses, but is commonly pursued by people who switch between diverse enterprises according to the current opportunities, rather than riding the booms and busts of a single industry. For example, bush food consolidators in central Australia actively buy from bush harvesters in good years, but switch to other jobs in bad years. However, they maintain the social capital of their networks for when a good year comes again.

- *The nomadic strategy* – a surprising number of people, Aboriginal and non-Aboriginal, are committed to the desert, but move around between jobs supporting Aboriginal settlements, mines, government offices or pastoral stations. They too need a good network of places to work at different times.

- *The 'exploitative' strategy* – a significant proportion of teachers, nurses, mining staff, bureaucrats, even researchers come to the desert for their 'hardship' posting for 2 or 3 years, to build their experience, their bank account or their reputation. They never intend to stay for long, but often make a useful contribution while they are there.

Also as for plants and animals, each strategy has its weak points (Chapter 3; Plate 6). For example, perverse policies can discourage investment in reserves by persistent small businesses, just as drought subsidies do for pastoralists (Box 12). Refuges can be damaged, such as when a tourism site is so over-visited that it loses its attractiveness, or a reliable but small water supply is over-exploited. Businesses must bring all the strategies to bear on the desert drivers – dealing with scarce and variable resources (whether grass or investment or people), a sparse and mobile population, remote markets and distance from the voice of government and head offices – and must be supported by government or company policies that do not expose each strategy's weak points.

However, the individual strategies are about making the best of the slice of the cake that is available. The most significant change in business strategy recently comes from trying to change the size of the slice, or even make the whole cake larger. This means building networks to achieve critical mass, and having a greater say in what financial resources stay in desert Australia. It is a parallel with the last plant strategy – the self-organising community. This is

when organisms get together to control and change the patterns of flows of resources. Plants do not do this consciously, of course, but humans do – and the key to this, in a remote and sparsely populated land, is communication and networking.

6.2 Using communications to deal with remoteness

The greatest advance for desert businesses of the past half century has, without doubt, been the communication revolution. From the message stick to the Overland Telegraph to the School of the Air and Royal Flying Doctor Service, Australians have always confronted distance isolation with the best technology available (see Chapter 8). Today, the internet, satellite telephones, mobile communications, broadband, email, teleconferencing, 'e-health', distance education and the increasing integration of all these services has brought about an interchange of ideas, information and business opportunity inconceivable a generation ago. For the first time, outback business people are meeting face to face in cyberspace while being physically thousands of kilometres apart: building trust, forming connections and contracting partnerships that are fuelling growth and opening up fresh horizons in their enterprises. This is gradually placing them on a more equal footing with the citizens of the world's megacities.

Subtly, but perhaps even more significantly to the human future, the new communications technologies are slowly erasing artificial boundaries drawn on maps by the colonial administrators of a bygone era, when the swiftest information flow between two points was the speed of a good horse. Business operators are seeing themselves no longer purely as Queenslanders, Territorians or West Australians, but more as outback operators, with much to gain from partnering with those once regarded as the opposition across the state border. The 'them' and 'us' of traditional outback business rivalry is eroding through greater contact and familiarity. As the world opens up to the communications revolution, the potential for outback businesses to form alliances with like-minded partners in lands and cultures around the globe heralds a profound change in both social and business attitudes and, potentially, in international and inter-cultural relationships. Rivals become customers or partners. Foreigners become friends.

Whip maker Mick Denigan, of Mick's Whips fame, was among the first outback business people to glimpse the power of the new technology to overcome the primary obstacle: isolation from markets. Remarkably, he became a pioneer internet trader, who was selling to a worldwide market at a time when he did not even own a telephone or have the power on at his home south of Darwin in the Northern Territory. Today, owners of his products include Bob Dylan, Bill Gates, the Duke of Edinburgh and several South American presidents.

In the mid-1990s, Denigan and his wife Shauna were running a thriving enterprise hand-making his trademark Australian stock whips and travelling around country shows and rodeos to market them. Demand from overseas visitors was growing, especially from the USA. 'We were doing well but we realised there was an even bigger demand for whips in the USA, and we were talking about doing the same thing there, travelling the American rodeo circuit. Then I heard about the internet and decided that would give us a presence in the USA without having to travel.'

'Where we lived, out of Darwin, we only had a "bag phone" [meaning that someone else took his phone messages and passed them on "in a bag"], but I knew Phil Hall of Aboriginal Fine Arts was marketing Aboriginal art overseas via the internet, and he kindly agreed to give me space on his website and to take orders.'

Denigan explains that part of the joy of running an outback business is the lifestyle. For a long time 'we ran the business out of the back of a truck. We could travel around Australia and still trade to the world market, collecting our orders wherever we stopped. We could be camped on Cape York, make whips all day then go fishing in the evening. When we turned up in Cairns a week or so later, a bunch of new orders would've come in'.

By 2007, the Denigans were running their station homestead as an eco-tourism venture, had five employees busy making whips and crocodile leather products, and were permanently online to the world with a real computer, phone and internet connection. After 12 years as a desert internet pioneer, Denigan concludes simply that, 'Anyone who runs an outback business is mad if they don't go online'.

6.3 The network feeling

While the communications revolution allows individual businesses to connect to the world, it also provides the platform for a second revolution in business practices: to tackle the effects of a lack of critical mass in the sparse population of the outback. Unlike the fragile, stand-alone enterprises of past decades, business today can be a fluid alliance of individual enterprises, some of whom may be hundreds, or even thousands, of kilometres apart – sharing ideas and working together in cyberspace to satisfy emerging customer demand for desert products and services across Australia and around the world. In many ways it is a model for the emerging global small business economy.

Using the internet, video-conferencing, web-logs, email and other communications methods to join hands and swap ideas across the inland, desert businesses are discovering new opportunities, partners and markets. Importantly, they are also overcoming local suspicion and rivalry, and working together to capture new markets.

The simple act of sharing ideas, comparing problems, offering advice and partnering with like-minded groups outside the local area turns erstwhile rivals into willing collaborators, who are ready to swap experiences and join hands with others in similar communities or ventures, according to Mike Crowe, who set up the Linked Business Networks Project for Desert Knowledge Australia (see Figure 10). In its first 2 years, this drew together no fewer than 330 outback businesses and 15 local business networks from five states and territories, into four nationwide alliances covering desert foods, mining services, tourism and outback construction.[153]

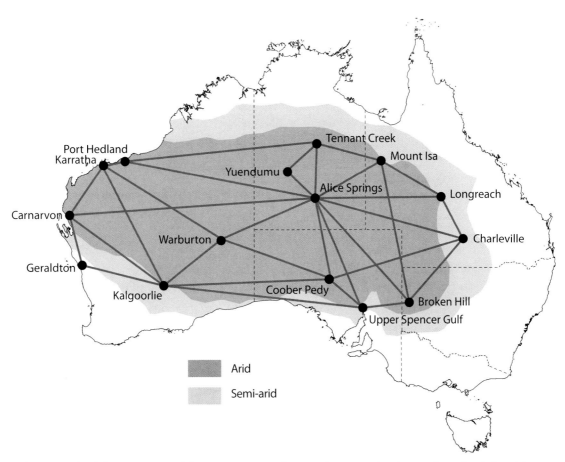

Figure 10: The network of desert towns with businesses currently actively engaged in the regular video links of the Cross Border Business Networking Project, or aspiring to this.[154]

The next phase, the Cross Border Business Networking Project (CBBNP), began in 2008, aiming to significantly strengthen the small enterprise sector in Australia's remote areas, so helping to build robust regional economies. The project establishes or supports existing regional industry networks of small businesses in nine regional centres across desert Australia.

'Has networking brought more business?' asks network development officer Joy Taylor. 'It's hard to know exactly what is due to the network when you're dealing with so many small enterprises – many of whom like to keep their cards close to their chest – but we're aware of several million dollars' worth of new business that came about as a result of our pilot networking project.'

The value of remote area business networks has also emerged in other ways, Taylor goes on to say.

'Some enterprises have used the network to band together to source supplies, so lowering their business input costs. Others have used it to carry out research into ways of doing business, which, as individuals, they would not have had the time or money to explore.'

One group pooled resources to investigate the practicalities of patenting new technologies, which, alone, they had avoided as time-consuming and daunting, illustrating a further dimension of the network – giving one another a collective push to get on with a task that most small business people would avoid (Box 14).

Another innovative undertaking was to stage the world's first 'virtual food fair', shipping 17 bush foods products to four locations around Australia where chefs, restaurant owners, caterers and other potential clients could find out how they were

BOX 14: MUTUAL SUPPORT TO GROW THE BUSINESS PROFILE[155]

Suzie Williams is co-owner of Polymer Fabrication in Kalgoorlie: a company that makes plastic pipes and products primarily for the mining industry. Williams joined the Mining Services Network with other mining services businesses from Kalgoorlie, Alice Springs and Broken Hill to use the Australian Innovation Festival to jointly promote the network members' innovative products, services and creative solutions.

She comments: 'These days, mining involves the movement of large amounts of slurries or saline water from one point to another across the mine site. The housings for the pumps are subject to severe corrosion and need replacing as frequently as every 3 months. So we invented a plastic housing, which is lighter and far more corrosion resistant.'

The prototype was still working fine after 7 years on trial at a local mine site, and Williams managed the process of patenting it herself, with assistance from some network members. She later shared her experiences with others, helping to demystify the patenting process. Polymer Fabrication became the first Western Australian regional business to reach the finals of the Department of Industry's Innovation Awards. This attracted considerable attention from would-be customers.

'They rang after the Award Night and we are now on the list of contractors whom they invite to tender. That's all we wanted – the chance to tender, no extra favours.

'We've found that once they meet us, see what we're capable of and that we are set up and skilled for remote area work, we get the job. Best of all we got back two lost contracts, as the winning [city-based] bidders didn't have what was needed to do the job! In both cases [changing to the other bidders] ended up being more costly for the mining company.'

produced, taste them and learn how to use them – direct from the business owners who talked to the four gatherings by video-conference (Box 15).

In other examples, enterprises have come together – both face to face and online – for training and skills, business development, mentoring and advice and problem solving. A benefit of this experience, Taylor says, is that business operators discover that others face the same challenge or hardship as themselves. 'Sometimes others have solved it, and can show you how. Sometimes it's a case of the businesses banding together to fix it. At other times, simply knowing you're not the only one up against the odds is what's important.'

'Developing new business and opening new markets is a key reason for networking businesses across the desert regions, of course, but it is far from the only reason. It's also about sharing ideas, reducing costs, getting advice from someone who's been there before, pooling resources, undertaking training and development and airing problems or concerns.'

But one size does not fit all: the different networks have evolved different clustering models to handle their individual opportunities appropriately.

6.3.1 Mining services alliance

Mining companies tend to put occasional big contracts out to tender, which are often beyond the capacity of individual local companies to take on. Here, to make the cake bigger, the key is for several enterprises to work together to capture the opportunities.

The Mining Services Network was formed to make links between a wide range of mining services businesses based across desert Australia. The network comprises over 70 businesses in five regional hubs at Kalgoorlie, Broken Hill, Spencer Gulf, Alice Springs and Mount Isa, with a diversity of capabilities and skills. What they had in common was a detailed knowledge of operating in remote and harsh environments, practical experience in the mining industry and established supply chains.

In Alice Springs, facilitator Mike Stellar watched with quiet amazement as a group of mutually suspicious small suppliers began talking about what they could do together instead of in competition. 'At the start one bloke stated flatly "I'm not sharing my business secrets with anyone!" but when the penny dropped that it wasn't about giving away secrets, but instead, working together in a complementary way, he became really keen', says Stellar.

The small companies in his group covered a range of services – power, water, equipment and communications – and it was not long before they decided to bid for major contracts collectively, which they could never have dreamed of servicing alone. As the minerals boom took root, the prospect of new mines opening up focused minds on the opportunities, and soon they were thinking about promoting the group's combined services Australia-wide.

An outstanding case was a $5.5 million mining services contract gained by five small engineering companies from Broken Hill in New South Wales, none of whom would have been able to tender for it on their own. Forming a consortium under the name Alliance Engineering Group gave them the scale, diversity and security to bid for, and win, the contract, said Operations Manager Joel Butcher: 'The initial contract was to overhaul a 1100 tonne dredge and separation plant for use at a mineral sands mine site in western New South Wales. Some of the companies had worked together before, but not on anything as big as this. One thing led to another and the new alliance was soon tendering for, and winning, projects as far afield as the Hunter Valley and South Australia, with a business plan to build a name for quality and efficiency among the big, listed mining companies.'

Small firms that once supplied only their local area made another discovery. Through the network, they found 'big brothers' who could introduce them to business opportunities thousands of kilometres away. In other cases, local support could help them through issues they were finding tough as isolated individuals (Box 14).

A centrepiece of the Mining Services Network's activities was a joint marketing effort, which included developing a network capability matrix and an online procurement centre. Fourteen businesses contributed time and resources to promote the network's collective products at the 2005 Goldfields Mining Expo. As a result, one member gained seven new clients and $100 000 of new business over the following 3 months.

6.3.2 Tourism alliance

Unlike the big contracts in the mining industry, tourism depends on many small opportunities – individual visitors or small parties choosing whether or not to come to your region and then whether to use your particular services. In this case, to make the cake bigger, networking needs to focus on marketing; the payoffs for individual businesses are harder to see in the short term.

Traditionally, the next town, or the town across the state border, has been viewed as competition by local tourism enterprises. The Tourism Network is helping businesses to overcome these age-old rivalries and join hands to create a richer experience for the growing flood of visitors to the outback.

The alliance started by linking together several existing local tourism networks and identifying special interest groups that would benefit from cross-border connections and regular online forums. In particular, the network aimed to cater for the growing drive-yourself adventure market, comprised of grey nomads, young explorers, family four-wheel-drive holidaymakers and international adventure seekers.

The tourism network includes a subgroup of Visitor Information Centres. These have helped each other in researching and testing different online booking systems. By pooling the findings, members saved time and money in finding a system that best met regional needs. The same members developed a calendar of desert events extending across state borders, which helps the Visitor Information Centre staff provide better information, planning and booking services for tourists in their onward travel.

Another subgroup of professional outback photographers used the Desert Knowledge Australia online forum as a business tool to establish their national network. Out of this, a local business cluster also formed in Kalgoorlie. A further subgroup formed to cater for tourism growth arising from the linking of the Ann Beadell and Outback Highways, sharing information about required permits, safety and the availability of services for tourists. Yet a third subgroup of businesses interested in food tourism joined the Desert Foods Network (see below) to participate in video conference sessions on regional food festivals and the possible creation of an outback-wide 'Bush Food Trail'.

Tourism illustrates a style of networking that is different from the mining network – smaller and more diverse activities in response to the smaller scale of individual opportunities.

6.3.3 Desert foods network

Networking is needed in the developing bush foods industry for a different reason – because any embryonic industry has limited capacity to develop its products and markets.[156] The unfortunate history of new industries in Australia shows that, too often, rivalries among small operators struggling to make a go of it end up tearing the nascent industry apart. However, the stimulating flavours of some central Australian bush foods are heading for home and restaurant dining tables thanks to a partnership between Aboriginal communities, scientists, food companies large and small, and supermarkets. Bush tomatoes, wattle seeds, desert limes, quandongs, sandalwood nuts, bush bananas and bush peppers are among the flavours of the desert now making their way into the twenty-first century cookbook.

Desert foods pose particular challenges seen in few other industries; their produce burgeons erratically across thousands of square kilometres, at the bidding of unpredictable seasons. The quality, shape, size, and colour of wild produce vary enormously, as does supply. Some are hand-gathered by local communities who are willing to share the piquancy, healthiness and sustainability of wild-harvested food with a wider market. Others are being domesticated for the first time in history – grown as crops by farmers, horticulturalists and Aboriginal communities to try to control the variability of supply. Besides being marketed through Woolworths and Coles, bush tomato products may now be found in airline dinners, in the Ghan railway's restaurant carriage and increasingly on gourmet shop shelves in the USA, United Kingdom and South-East Asia.

According to bush foods scientist, Maarten Ryder, 'a vital part of this development is the involvement of Aboriginal communities in a new industry, and how it can generate new jobs and enterprises in remote areas'. The bush foods industry marries the traditional knowledge of these long-term desert dwellers with modern scientific approaches, such as the breeding of new domesticated varieties that can be farmed in the low rainfall regions of the inland. An early example of quick success from this partnership was to work out that insect damage on bush tomatoes was being caused by a weevil that managed to get in while they were in storage rather than coming from the original harvest, which enabled easy control to reduce losses.

The intellectual property in any new crops produced remains with the traditional owners, who thus have a say about how the bush food industry develops. At the same time, researchers are helping Aboriginal communities to catalogue and conserve the wild plants that have been the basis of their way of life for millennia.

The Desert Food Network brings together many small and remote desert businesses, including bush product wholesalers, specialist retailers, primary producers, medicinal plant growers, caterers, restaurants with interests in wild food cuisine, and mine site rehabilitation businesses. Its activities have included teleconferences on markets, professional development sessions for chefs and food event organisers, collaborative marketing of member products at regional events, the virtual Desert Food Fair (Box 15),

BOX 15: WORLD-FIRST VIRTUAL FOOD FAIR[157]

Jason King is proprietor of Bell's Milk Bar in Broken Hill in New South Wales: an authentic 1950s milk bar, known for its original malted milks and soda spiders. They use syrups and cordials, which are hand-made on the premises just as they were in the 1950s, and follow a tradition extending back to 1892 when the shop was established.

King joined the Bush Products and Local Foods Network in Broken Hill. At a cross-border video-conference on food festivals and local food events, the idea emerged to hold an outback-wide Desert Food Fair – with a difference. Instead of selecting a location to hold a trade fair and then having to find the time and money to attend, network members decided to organise what is thought to be the world's first virtual trade fair, using video-conferencing technology.

King helped organise the event in which all network members were invited to participate. A tasting menu was drawn up and bush dukkas (bush seed dips), preserves, coffee blends and syrups were sent to sites in four Australian states, along with serving instructions. On the day of the Fair, 13 products were demonstrated, discussed and sampled.

King demonstrated how to make his milkshakes and soda spiders – meanwhile, participants sampled milkshakes and spiders that were made at other sites following his instructions. Afterwards, he made sales to people who had participated, to people they had talked with, and to customers who visited a website promoting the products in the virtual trade fair.

'Using the feedback obtained at the virtual trade fair, I was able to successfully pitch the syrups and cordials business idea to a panel of high-profile judges at the recent NSW Young Bizstar Competition in Sydney', he says. 'Winning the regional category lifted the profile of Bells Syrups and Cordials beyond my expectations, and gained me a $12 000 business support package. It was a great feeling to stand in a city office and sell the idea of running a successful manufacturing business in the outback.'

an Alice Springs wild foods tourism trail and the 'Blue Sky Red Earth' bush food festival at Broken Hill. Two network members also attended Restaurant 08 in Sydney on behalf of the whole network. One of the exciting ideas to come from the network is a Bush Foods Trail around Australia: a gourmet tourism adventure enabling visitors to sample unique products and cuisine as they tour the outback.

'There's great potential for bush tucker as part of tourism in Alice Springs and across the desert', says Raylene Brown of the Aboriginal business *Kungkas Can Cook* in Alice Springs. 'We joined the network to have a voice in that, and to learn about what businesses in other regions are doing. I've met wholesalers, growers, harvesters, producers, chefs and cooks and learned a lot to help our business.'

6.4 The outback arts boom

Given the remoteness of markets, outback businesses need to pay special attention to those products in which they have a strong competitive advantage – that is, products that nobody from elsewhere can step in and steal. Services that have to be delivered locally, such as haircuts and car repairs, are safe from the forces of globalisation, but these services are only in demand if someone is living in the outback. The underlying business opportunities are mostly based on the natural and cultural capital of the desert – the rocky ranges, the mineral resources and the local culture that you just cannot find anywhere else on Earth. Given a business based on one of these, then all the lessons we have been talking about so far need to be applied.

The pre-eminent example is Aboriginal art. This has, in less than a quarter of a century, evolved from a cottage industry to one worth an estimated $500 million a year, marketing the products of more than 100 community-owned arts centres to an enthusiastic world market. It is not unknown for a single auction of Aboriginal paintings to gross more than $10 million.

Founded on the 40 000 year old tradition of what has been called the world's oldest living culture, Aboriginal art has flowered in the twenty-first century into an astonishing diversity of painting styles and products – ceramics, textiles, sculptures and carvings, dance, music, theatre, cuisine and tourist gifts. It began with very traditional dot painting styles drawing directly on how people used to represent stories in river bed sands with rocks and feathers and colours, and evolved to brilliant modern acrylics with representational animals, people and homes, as well as Albert Namatjira's watercolours with their extraordinary ability to capture the colour resonances of central Australia. Taking full advantage of modern communication technology – the internet, email and even virtual reality – isolated desert communities now market their Aboriginal art direct to customers in the United States, Germany and many other countries worldwide. In this respect, Australian Aboriginal peoples, together with the Inuit of Canada and the Maori of New Zealand, are world leaders in a fast-growing movement to share the creations and concepts of indigenous cultures with a global audience.

Although Aboriginal art had been sold to passing visitors since the late nineteenth century, it only began to flourish with the advent of the community-owned arts centres. This movement grew from two strands – 'mission art', an activity encouraged by a handful of religious missions from the 1940s on, and the Papunya art movement. In the Papunya movement, an invitation to community elders to share their knowledge of country and wisdom with school children through art led to a wildfire of creative expression, starting in the 1970s, and laying the foundations for today's thriving enterprises.

Today, that art is online and being sold over the internet from places as remote as Yuendemu and Warburton. Desart – the Association of Central Australian Aboriginal Art and Craft Centres – networks these isolated centres together, to help provide continuity of marketing to the outside world (Figure 11). Art is a classic example of a high-value product that is relatively cheap to transport, so it can be competitive from very remote locations. The high value, of course, is based on the cultural content.

John Oster, Executive Officer of Desart, says that for an increasing number of buyers, desert art is less about acquiring a fabulous artwork or an investment and more about learning about the spiritual life and engagement with the desert landscape that inspires its creators. In addition to the fast-ramifying contemporary art movement with its plethora of new styles and materials, there is also the 40 000 year heritage of Australian rock painting: the greatest such collection on Earth. This, he says, is still being added to or renewed out of deep cultural knowledge and, although it is scarcely portable, it is also being reproduced in many modern paintings, carvings and artworks, as well as being photographed.

The advent of modern technologies has not only broadened the range of Aboriginal artistic media and scope for expression, but technologies such as data storage are enabling the art to be imaged, catalogued and stored – safely and forever. This not only preserves the entire life work of individual artists and whole communities, but also provides a guarantee to the buyer that the work is genuine, as well as a wealth of detail about its origins and significance. Such actions, in turn, help to assure the industry's competitive advantage.

Thus, in a remarkable fashion, cultural, ecological and spiritual associations with the desert landscape from the deepest time – thousands of years before 'civilisation' – are gaining fresh significance to a worldwide audience of admirers, through the portal of the most modern of technologies. The art is displayed in leading galleries in New York, Paris, London, Singapore and Beijing, as well as online everywhere. These outback artists are influencing

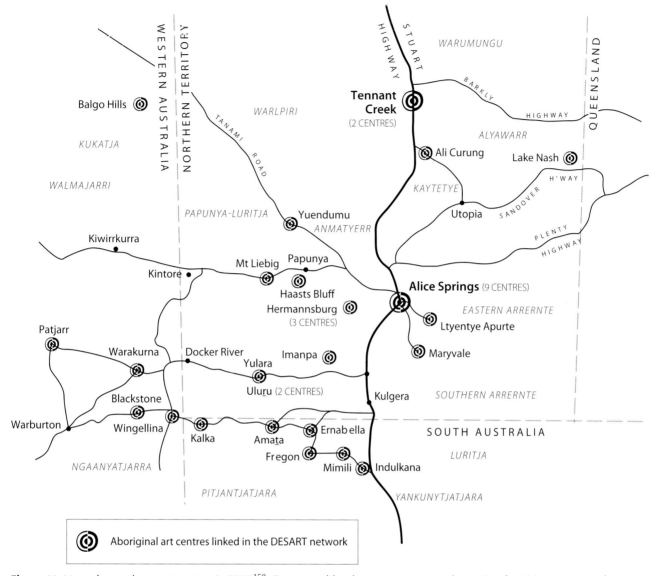

Figure 11: Map of most desert art centres in 2008[158]; Desart enables the centres to network to gain advertising power, and to some extent substitute for each other if one community is out of action for a while, thus ensuring continuity of supply to the outside world in a socially variable environment.

human artistic expression worldwide, achieving the goal of Valerie Napaljarri Martin, a previous Chairperson of Desart, 'that everyone can know about us, so we can carry on, so our kids can carry on forever, even when we're gone'.[159]

6.5 Telenomadism

Besides those businesses that are competitive in the outback because their products are unavailable elsewhere, there are also some that can function from anywhere. If these business people want to live in the outback with an outback lifestyle, well, they can! The era of the 'grey nomad' – the retiree who loads up a caravan and four-wheel-drive and sets off in search of outback adventure – is heralding a new style of mobile enterprise.

With the internet now accessible by wireless, satellite or phone in many places in the outback, the nomadic trader no longer needs to leave their company or share portfolio in the hands of another while they roam (Plate 7d). An increasing number take their laptop, blackberry and investment portfolio with them, to maximise income in retirement or keep the business ticking over while taking the adventure holiday of a lifetime.

The tools of the modern business executive – normally associated with cities, suits, aircraft and a high-pressure lifestyle – are equally serviceable in the hands of the tieless, unshaven wanderer in the Earth's more remote regions (though a good connection may not yet be guaranteed). Just as the younger generation has discovered the lifestyle joys of telecommuting from home or combining a web business with child-raising, the older generation is discovering that business can be done amid desert surroundings that are far more inspiring and congenial than any city office.

When Keith and Trish Bashford took to the unsealed red roads of the outback for a year-long retirement tour, the plan was to get away from it all. On the way, they found that they needed to manage their shares and, after a while, doing a bit of business on the stock exchange from the most remote of locations seemed second nature. 'We met lots of people who were funding their retirement doing a bit of trading or running their self-managed super fund while out on the track. It's really extraordinary how sophisticated people were able to be: using the technology and the better investor websites, while roaming the bush. Nowadays there's almost no difference running a portfolio from a fishing camp on the Drysdale River (in northern WA) by satellite phone, to doing it from Pitt Street', Bashford says.

A more traditional grey nomad activity is to stretch out the finances with a bit of fruit picking and casual work. In an original twist, the Queensland town of Barcaldine decided to take advantage of the tide of retiree skills and talent that sweeps through in camper trailers each year. Barcaldine Mayor Rob Chandler said the Shire has developed a pilot program to employ retirees on public works, tourist attractions and amenities in need of a bit of volunteer labour or restoration: 'They're all looking for something to do. The workforce does not want

them any more and in general people are bored when they retire. They want to use their experiences and their skills – and what better way to do it than in a voluntary capacity in some of these small regional towns.' With the rapidity characteristic of the bush telegraph, the idea has spread quickly to other towns such as Winton in Queensland and Roxby Downs in South Australia, helped by advice from Barcaldine. This all points to a rich new vein of labour, skills and talent that desert areas, both in Australia and around the world, can tap – potentially a mobile international labour exchange.

6.6 Business lessons from the deserts

The desert environment will punish any business that does not make wise use of resources in ever-changing and unpredictable circumstances. Like the plants, animals and nutrients of the deserts, the resources that a successful desert business depends on are sparse and it must have strategies for surviving a prolonged 'dry spell'.

As we have seen in Chapter 3, sometimes these strategies consist of putting down deep roots to tap resources inaccessible to other businesses, as for the coolibah or mulga. Sometimes, like the outback's nomadic birds, it means being able to wing swiftly from one transient opportunity to the next. Often, as for the knob-tailed gecko, it consists of storing resources (or money) against the hard times or, like the ephemeral paper daisy, expanding business rapidly when conditions are favourable and contracting just as fast when they are not. Sometimes, businesses must mimic fire-loving plants: spotting a vacant niche in a landscape where a bushfire has scorched away the competition. And, sometimes, like the termite, it consists of the ability to prosper through specialised persistence on material apparently so unpromising that no other creature can take advantage of it. The desert is a good business teacher because the five basic strategies for desert survival are much the same as those for individual businesses. Supportive governments and communities need to ensure they do not trigger the weak points of the strategies – discouraging the wise storing of reserves, destroying the attraction of critical refuges or damaging the connections between places.

In the twenty-first century, however, the communication revolution and the expanding networks have added a fresh dimension. The desert business is no longer isolated and dependent on its own resources and skills – it can draw on customers, suppliers and partners who may be thousands of kilometres away – even in another country. It can communicate at light speed with a buyer in New York or a seller in Amsterdam, a partner in Kalgoorlie or a business angel in Julia Creek. Because of their awareness of the significance of distance, desert businesses are on the front foot in taking this business model to the world, especially among small enterprises. These strategies are like the self-organising communities of plants and animals: partnering up to do better together than they can alone.

Critically, governments can support such linkages with good communications systems and business support for networking, or they can make this difficult and cause the business efforts to fail. For example, the Australian National University's Howard Morphy argues that well-meaning interventions to control the Aboriginal arts industry can undermine its independent success, and that it is better to let those with long local experience in the industry develop their own solutions. On the other hand, as he also points out, once the parallels are recognised and understood, other industries such as remote tourism or bush foods and medicines can learn from the successes of the arts industry.[160]

Throughout history, while governments and peoples have warred, wrangled and disputed, businesses have traded. While the three players are inextricably entwined, it is usually in the interests of business to have peace, unfettered commerce and mutual understanding. Large multinational businesses are capitalising on the era of globalised commerce and communication and have dominated the trade scenery much as the dinosaurs did the Jurassic era. Every few weeks, it seems, another global monster is dismembered or devoured by its competitors, leaving ever more titanic and impersonal corporations, which resemble countries more than business

ventures. But between their great feet, amid the undergrowth, lithe and agile competitors are stirring: the mammals of this analogy are the globalised small enterprises that are exploiting both the cutting edge of cheap communications technology and their ability to form loose and versatile networks with one another across the planet.

The globalised small business has the opportunity to go to the world market without losing its identity, without sacrificing its friendly face and personal touch, its loyal customers, its ideals, its family character or its community values. Through technology and networks, it can now harvest the skills, resources and contacts it needs to prosper internationally. For the first time, the street outside the local store leads right around the world.

The global small business can be run from almost anywhere with access to modern communication – even from the back of a truck or camper van. It is transforming the nature of isolation and resolving arguments about the viability of remote settlements. With the web of communications that increasingly spans the Earth, nowhere will be truly isolated and distance from markets is no longer a barrier, especially for creative knowledge-based products.

The opportunities for business synergy have barely begun to emerge as the tendrils of advanced communications technology begin to penetrate the Earth's most remote regions. The knowledge required to grow food and build sustainably, use power and water sparingly, manage landscapes, create art, leisure and pleasure and govern communities amid dry conditions is not a commodity of much interest to large corporations. It is too specialised, and requires too much fine-tuning. It is, by its very nature, the sphere of small enterprise.

Another significant implication of interconnecting small businesses and individuals across vast distances is the gradual dissolution of ignorance, fear and suspicion of 'the other'. Just as outback ventures have found that their competitors across the state border turned out to be allies in a mutually beneficial business expansion, small enterprises that interconnect around the world will play a powerful role in positively influencing the views of their communities towards those who live in other lands and cultures. The reduction in global suspicion and mistrust among ordinary citizens, though unquantifiable, must play a part in reducing tensions and the scope for misunderstanding and conflict. From the time of Marco Polo, or even the Phoenicians, small merchants and family companies have been the bridges between cultures. In the twenty-first century, with growing opportunities to join hands with almost any equivalent small business on Earth, they can be harbingers of understanding and shared benefit at a time when governments stand divided.

The fact that Australia needs to make these approaches work within its desert region offers the country two profound opportunities – to be a world leader in adapting to the dry times ahead, and to help further the revolution that will interconnect small ventures worldwide – heralding a global small business sunrise that is resilient to economic recession, and enhancing understanding between the world's ordinary citizens.

7

Surprising settlements

The bulldozers rolled down the main street of the old town, erasing the last remnants of the place where several hundred pioneers had lived and laughed and loved for the past 50 years. In the space of two years, they had all moved out and the town's final destruction had begun.

But this was no act of wanton vandalism in a slowly declining neighbourhood, it was not a desperately fought slum clearance in a Brazilian favela, nor a careless act of war. This was the old South Australian mining town of Leigh Creek, and its useful buildings had been sold at auction a few years before. Its inhabitants had moved cheerfully into new (and better) quarters at Leigh Creek South, 13 kilometres away, and the bulldozers were initiating precision cuts to unearth the major coal seam that had been found to underlie the old, company owned town.[161]

Actually, the destruction of the old Leigh Creek was not quite as dramatic as this – it took a few years for the mining operation to move in. Nonetheless, the re-location of a whole town for business purposes would be an alien idea for people who inhabit the permanence of today's cities: Leigh Creek illustrates just one of the ways in which desert settlements are so different from those that most of us live in.

On a drive through the outback towns of Yulara, Halls Creek, Yuendemu or Cue, the streets and houses and people and roads all bear a superficial similarity to those of a city suburb, so one can be beguiled into assuming that outback towns work the same way as other urban settlements. But they do not. Sometimes they just pack up and move on. Sometimes their inhabitants scatter. Sometimes they are spread out so thinly that you cannot even see there is a settlement there. In many ways they are unusual, and treating them as if they were ordinary towns leads to some serious mistakes. Understanding these settlements is vital to managing the outback.

7.1 Desert settlements are different!

Leigh Creek is a mining town in the Flinders Ranges of northern South Australia. It was founded in 1941 by the South Australian Government's Engineering and Water Supply Department to exploit the nearby coal fields and reduce the state's dependency on imported coal. The Department built the town beside the railway to house its workers, and got to work digging. Some years afterwards, they discovered to their dismay that the main coal seams ran right under the town itself. So they rebuilt the town 13 kilometres away, where it stands today. In 1980, they moved all the inhabitants across, sold off the old town – including the police station, school, and even the morgue – and started mining where it had once stood.[162]

The new township, actually called Leigh Creek South because of its move, has better facilities than the old one, including a little shopping centre, supermarket, service station, post office, sports fields and, of course, a pub. But the new township is all owned by a private company, Flinders Power, so there are no real estate agents and people cannot just move in. Visitors can drop in to the shops, but it is really a private settlement. If another seam were to be found under the new town (although they did check!), you can be sure they would demolish it too, and move again. This is a town owned and managed by one company, which has little hesitation in picking it up and moving it if it is in the wrong place.

The ghosts of ephemeral mining towns are common. In wetter areas, mining towns often live on after the mines have gone because an alternative local economy has developed. In the desert, though, mining towns only occasionally manage to meta-morphose into a service centre (such as Kalgoorlie, Broken Hill, and Mount Isa) or re-focus on some other livelihood (such as Coober Pedy diversifying into tourism). Historian Barry McGowan is an enthusiast on this topic. In his book, Ghost Towns of Australia[163], he documents literally hundreds of mining settlements – some long dead such as Arl-tunga east of Alice Springs, others gracefully decay-ing today, such as Cue and Big Bell. Only a few survive into another life. Deserted ghost towns abound in other deserts such as the Atacama too – there are many decaying rows of workers' quarters scattered through the hinterland of the modern copper mining town of Calama. Like an ephemeral desert paper daisy, these sorts of outback towns just pop up for a while when resources abound, and vanish again when they run out (Plate 6b).

Yulara (Ayers Rock Resort) is another single-purpose town, which is focused on the tourist delights of Uluṟu-Kata Tjuṯa National Park. Yulara, like Leigh Creek, is mostly privately owned (now by Voyagers Hotel Group). People cannot really live there unless they are contracted to the resort. The difference between Yulara and Leigh Creek is that Yulara is never likely to be moved, because it is not dependent on an exhaustible resource like a mine, but on the permanence of Uluṟu (Ayers Rock). Yulara is more like the refuge-dwelling cycad palm of central Australia: focused on a rich reliable resource that needs to be tended and protected – in this case the environmental and cultural tourism attractions of Uluṟu.

The most widespread form of outback settle-ment is the pattern of 4000 pastoral stations spread thinly across three-quarters of the continent, each home to one or a few families. The Barkly Table-lands in the Northern Territory and the Murchison Shire in Western Australia bring home the fact that pastoral settlement consists mainly of these spread-out homesteads, because neither the Barklys nor the Murchison have any gazetted towns at all. The Murchison Shire headquarters is Murchison Settle-ment, which was established with a little shire office and a roadhouse in the 1990s. Now with a population of 16, it is no bigger than any one of the 29 vast pastoral properties it services.[164] The Bark-lys have no such centre, though some company-run properties have relatively large populations, such as the Australian Agricultural Company's Brunette Downs with about 50 people (and 70 000 cattle). The police station at nearby Avon Downs Station serv-ices 147 000 square kilometres![165] In regions like these, pastoral settlements are scattered sparsely over hundreds of thousands of square kilometres, yet still see themselves as a pastoral community that often gets together for social occasions such as race meetings and Landcare events. A regional pastoral settlement thus resembles a long-lived perennial mulga tree: spreading its roots across a wide area to harvest sparse resources and survive through variable seasons.

Aboriginal settlements are also distinctive – but in another way. Notwithstanding the degree to which Aboriginal people have been encouraged (sometimes forced) to settle in particular centres over the past 150 years, desert people are still highly mobile. The population of the small community of Engawala in the Northern Territory, for example, went from the average of about 135 people to only two for several weeks at one stage and then ballooned out to over 200 for 'sorry business' – mourning after a resident's death – at another time.[166] People were moving around from one place to another for extended periods for sport, for cultural reasons, for work and to access a hospital – at times to get away from family and at others to interact socially. This is a settlement where the human community is not even tightly bound to a single location in the way most urban dwellers would think was 'normal', but wanders freely as social and other needs decree. Such Aboriginal settlements are also generally not public, but are formally owned by a community of people, rather than a single private concern as with the mining or tourism towns. This leads to different dynamics when it comes to agreeing on services, their standards, and how to deliver them – and the principles used in suburban Australia break down. The way that some Aboriginal people use their settlements is more like the desert nomad: moving in groups from one place to another as priorities and resources change. Indeed, most desert people worldwide used to obey this basic rule of mobility in order not to overtax local resources.

Desert Australia also has an important type of larger settlement, the service centre – cities such as Alice Springs, Kalgoorlie, Broken Hill or Mount Isa. Although mining may be the mainstay of Kalgoorlie and Mount Isa, these have evolved into more complex communities serving a range of local needs. These changes are most obvious in Alice Springs, which has no single dominant industry driver, such as mining. Rather its prosperity depends on tourism, defence and delivery of government services, and as a transport hub, serving governments, industries, outlying settlements and all the people who live within a few hundred kilometres of the town itself. More than anything, towns such as Alice Springs depend on the people who live and work in their hinterlands. Without

those, most of the economy of Alice Springs would fade away.

To appreciate how distinct the needs for town planning and management are, imagine if Sydney was planned to last only for 20 years, like many mining towns: its design would have to be radically different. If Melbourne City Council had to set up their bus, water and library services to cater for a population of 3 million this month, but only 30 000 next month, it would need some very different and highly flexible approaches to providing these services. If Brisbane Water had to deliver water to a million consumers spread across the whole of outback Queensland instead of just the focused area of the city, it would need completely different infrastructure (and a great deal of piping!). These examples illustrate how desert settlements would look if they were scaled up to the size of the cities, where most of Australia's planning regulations, local government processes and service technologies are shaped. It is clear that traditional urban models of services based on fixed, public settlements will often fail to meet the needs of people who live in the desert and obey its distinctive rules for living – Aboriginal or non-Aboriginal. To understand how to respond to this, we need to explore the drivers of these various types of remote settlement a bit further.

7.2 Communities of livelihood

Desert settlements are small, and most of them are based on a single type of livelihood. Their population is actually a 'community of livelihood' – a group of people almost all of whom have a common interest in essentially the same type of activity, whether that is mining, tourism, grazing, conservation management or cultural affairs. Even people in a service support role are interested in the primary activity, such as mining, because without it they would not have any customers to service. This is unlike a city, where they would simply service someone else.

The outback *is* the outback because it is much less productive per hectare than the closer settled areas of Australia, notwithstanding some intensely valuable nuggets of activity around mines. Consequently, almost all desert settlements, Aboriginal or not, are located in areas with fewer livelihood options than regions of closer settlement. In general, people live there for one of three reasons, which are by no means exclusive: (i) because of commitment to place (mostly cultural, particularly Aboriginal, but also non-Aboriginal); (ii) because of a place-based business or livelihood opportunity (e.g. tourism, mining, pastoralism, nature conservation, etc.); or (iii) to support (family, partner, etc.) and service (govern, teach, doctor, build, massage, feed, etc.) those who are there for one of the previous two reasons (thus creating some of the business opportunities).

By 'place-based' business or livelihood opportunities, we mean ones that depend on being there in desert Australia – gold cannot be dug up from anywhere except where it is located (though it can then be processed elsewhere); tourists can only see the MacDonnell Ranges in central Australia (though they could see other red ranges in Pakistan or Arizona, and their money often ends up at tour companies' city-based head offices rather than in the desert); and fire management in the Tanami Desert can only be carried out *in* the Tanami. Businesses that can exploit desert Australia's resources, or service their people from afar, are increasingly unlikely to locate themselves in the desert; instead they act on a fly-in-fly-out (e.g. mining and government services) or remote (e.g. internet) basis. Only the businesses that require people there on the ground are resistant to this centripetal attraction away from outlying small settlements caused by globalisation.

Most people living in desert settlements therefore ultimately depend for a material living from some mix of six sources (Table 3): natural resources, driven by climate, which underpin grazing and tourism, as well as customary harvest and bush foods; minerals, which are naturally patchy, but also affected by volatile markets; cultural resources, including Aboriginal and outback cultures, but also social networks that help support people; public payments that underpin welfare as a matter of citizens' rights; and payments for services, either in the national interest, or to support people living in the region. These categories are repeated from the smallest to the largest

settlements, albeit with very different balances depending on size and the settlement context.

These create *communities of livelihood* – that is, people who are bound together by virtue of depending on a similar resource – for each of the sources above, including pastoralists, tourism operators, miners, Aboriginal family networks, people on the dole, conservation rangers or small businesses providing goods and services in town (though the last group, like government administration, ultimately only exists to service others).

It is not usual to include livelihoods based on cultural relationships and social capital as one of these categories, but this is clearly so for Aboriginal communities, and is indeed significant for others, such as pastoralists. Aboriginal people depend on these connections between family members in different settlements. Pastoralists obtain significant support from their neighbours in times of hardship. Indeed, a more familiar example of a 'culturally based livelihood' is that of a priest or pastor, whose livelihood depends on servicing people's spiritual needs – not so different from Aboriginal elders sustaining their dreaming *tjurkurpa* on behalf of their community.

In some cases, this cultural and social capital can be converted into more conventional livelihoods through cultural tourism, arts and crafts. At other times, it may be a source of mutual support in times of need, such as during droughts when pastoralists call on neighbours for help, or during political uncertainty when Aboriginal people move into town to call on relatives there. We can call all these emergent values '*cultural services*', by analogy with ecosystem services.

It is easy to see that, in desert Australia, the underlying nature of most of these sources of livelihoods is variable, unpredictable and outside local control (Table 3). This is obvious for plant growth that is driven by the climate, as well as for the market value of minerals, which determines whether mines keep operating. It is also true for welfare and for servicing the public interest, which depend on government investment. This is variable and unpredictable because of the vagaries of politics;

Table 3. Main types of resources on which people base their livelihoods in desert Australia, the livelihoods, and how variable these are.

Resource supply for livelihood	Livelihood examples	Level and source of variability
Renewable natural resources of the desert (e.g. forage for stock, bush foods, other bush products, environment for nature tourism)	Pastoralism, tourism, bush food industries, subsistence harvest[a]	Highly variable in time (and space), dependent mainly on climate (though some products, such as specialist timbers and tourism, are less sensitive to this than others such as grazing or bush tucker)
Non-renewable natural resources (e.g. minerals, oil and gas)	Mining	Highly variable in space and time – minerals are very localised and often only last 10–20 years; their value is also highly variable depending on distant world market
Cultural resources and social capital (Aboriginal culture, pastoral culture, outback ethos, and social capital from community networks)	Maintenance of culture, church, family contributions, community support – 'cultural services'	Potentially variable, but mostly dependent on local efforts to maintain relationships, and hence mostly under local control
Public transfer payments	Welfare	Variable over time: should be stable, but subject to distant policy decisions, with significant changes in recent years
Payment for servicing the public interest more or less independently of the resident population (e.g. conservation management, but also specialist activities such as defence and quarantine)	Conservation management, defence, quarantine	Variable over time: more or less independent of the size of the resident population, but subject to distant political decisions about priorities
Payment for servicing the resident population's needs	Small businesses in the service sector (e.g. food, hair cuts, small business advice, etc.), all tiers of government/ administration, and related authorities and businesses	Variable over time: dependent on the size of the population that is resident for any of the previous reasons

[a] Solar energy from the sun and geothermal energy from the Earth's crust could be included in here as renewable resources that are *not* variable, but, as discussed in Chapter 8, these opportunities have hardly been capitalised upon as yet.

in particular, the level of investment is set by political forces mostly located outside the desert region, and allocated according to priorities largely outside the influence of desert dwellers themselves. Servicing the needs of other residents is obviously very dependent on the population level, and hence the vitality, of the other forms of input, although it is buffered by government spending, which has inertia and does not usually change as fast as the climate does. Nonetheless, if there is a trend of declining population as a result of the other forces, demand for services will soon track that decline.

In fact, it is cultural resources and the social capital that underpins cultural services that are among the least variable sources of livelihood and those most under local control.

In short, the livelihoods of people in desert Australia are dominated by activities that depend on sparse or patchy, variable and unpredictable drivers, whether biophysical (such as climate) or social (such as distant policy), or which respond to

these indirectly. All this occurs in some elements of any economy, but by definition it does not dominate the livelihoods of most of the world's population centres. It follows that settlements and communities in desert Australia – and indeed in remote regions worldwide – will wish, and need, to arrange themselves in ways that are different from cities in order to respond to these causes of uncertainty.

This is not a new problem for desert dwellers, of course, because plants and animals, as well as pre-colonisation Aboriginal people, have come to terms with exactly the same patterns of uncertain resource inputs, and have spent many millennia adapting to them. Most desert people have to deal with substantial uncertainty in the things that they depend on for making a living. Combined with the effects of small populations, it is not surprising that the settlements might behave in ways that echo the life history strategies of desert biota (Chapter 3). Some of them are ephemeral, some are refuge dwellers – depending on rich hotspots – others are persistent perennials with good resource harvesting structures, and still others are nomadic.

We should not take the biomimicry parallels too far, but we can also expect the behaviour of these settlements to exhibit the same types of weak points as the plants' strategies (Box 7; Plate 6). Thus, persistent communities need to build up financial or social reserves in good times, 'nomadic' communities must have the resources and links to be mobile between settlement locations, dependent communities must have the support of a 'big brother' settlement (such as a nearby mining or tourism centre) and refuge communities depend on the maintenance of their 'refuge' (such as the tourism attractiveness of Uluṟu). As with plants, a failure to understand how the different community strategies operate can disrupt the weak points and create dysfunctional settlements.

7.3 Aspirations and trade-offs

Imagine being a pastoralist living on an isolated station with your family and perhaps an offsider; would you want a good power supply or a football oval? Or, if you were single miner, flying in for 2 weeks at a time – would a good internet connection or a kids' playground be uppermost in your mind? Or you run a peaceful remote tourist stop-over – do you want a plasma television and a large power generator or a decent road through the nearby creek so you do not get flooded in? City dwellers would probably think that all these items are necessities and, even if someone did not want one of them, a significant number of other people would, because there is such a diversity of people in a large city. By contrast, inhabitants of the unusual desert settlements – based on narrow communities of livelihood – may have quite a different a way of setting their aspirations.

In the long term, settlements only thrive if their aspirations for services can be met at a reasonable cost. The problem is that remote communities have rather different aspirations for the services in their settlements than in most coastal cities or towns.

One may think of three broad categories of aspirations for services, corresponding to the breadth of consensus that may be had about them. Firstly, there are those that are regarded as a necessary right in our society at large; secondly, that local communities regard generally as the norm; and, thirdly, that are essentially a matter of individual choice.

In the first category, there is some base level of health, education, sanitation and communications that would be regarded as a non-negotiable necessity for living in Australia, which is imposed as a social norm whatever individual communities might think. For example, basic education is compulsory, cars require registration, violence to women or children is not acceptable and people must have water to drink. In the third category, there is a substantial zone of discretionary aspirations that individuals choose for themselves – whether access to a bowling alley or a golf course is more important, a mini or a four-wheel-drive, an old telephone or Voice-over-IP on high-speed broadband. The last example illustrates how standards change over time; just 5 years ago only enthusiastic geeks would have regarded broadband as vital, but today Australia is close to requiring it as a non-negotiable necessity for modern life. It is easy to think of similar examples from the past that have evolved from option to necessity – access to electricity, inside toilets and seat belts in cars. In-between these extremes lies the second category: the values and norms of individual communities, with regard to relatively major issues such as local law and order, recycling or planning standards for consistent architecture.

In large settlements containing many communities (as in cities or large towns), there is a great diversity of aspirations, and the three tiers of governments (local, state and national) essentially work out a consensus for the standards of education, sanitation, infrastructure, communications and so on, at their respective scales. So, it becomes common for our levels of government to be structured to devise and maintain such standards. However, as we have seen, small communities often have dominant norms that differ from the national average. This is evident in the dispersed pastoral community – people who are happy to live remotely and cope self-reliantly, with poorer services and distant medical help in exchange for a lifestyle and a particular livelihood opportunity. It is even more evident in mobile Aboriginal communities for whom cultural attachment to country is far more important than easy access to the Adelaide theatre. The homogeneous views of communities of livelihood like these are usually lost in a large settlement, or occasionally dealt with as special cases.

Mark Moran and fellow researchers at the Centre for Appropriate Technology have looked at desert communities in terms of the 'five capitals' (see Box 16): financial capital; physical or built capital such as infrastructure; natural capital in the environment; human capital; and social capital.[167] They observe that different communities have quite different expectations of the state of their different capitals, and are happy to trade these off; for example, a community may be happy to have higher amounts of natural capital, but less physical capital than people elsewhere. Figure 12 shows this schematically – although cities are very high in financial, human and physical capitals, they may be relatively lower in social capital and much more poorly off in terms of their environments compared with a remote pastoral station that is poor in financial and physical capital, but much better off for natural and perhaps social capital. Many pastoralists talk of making this trade-off for the sake of the lifestyle. Mining communities may be happy to accept some privations knowing that they are simply flying in and out, or are making loads of money anyway and will return to the city. Likewise, remote Aboriginal communities that may seem low in financial and physical capitals may be very high in the other capitals when functioning well, and this may be an acceptable trade off for living close to ancestral lands.

Of course, if the human and social capital drops away as well, as has happened in terms of health, education and family violence indicators on some remote communities today (Figure 12d), then the trade-off may cease to be attractive. In fact, when the levels reach the minimum below which the Australian community would regard the situation as intolerable (illustrated in Figure 12 by the inner black pentagon, which applies to all the capitals),

intervention by government to restore minimum standards may be warranted.

The point about all this is that the communities living in small desert settlements tend to have a much narrower range of aspirations than that found in larger settlements. Desert dwellers usually have very clear ideas of their own about what they need and how it should be delivered. In part, the ideas also differ among desert settlements according to the strategies of their communities of livelihood, and the weak points of those strategies. And there is often a mismatch between these ideas and those of the distant bureaucrats who are charged with delivering the service according to a particular model or policy. Simply assuming that all communities aspire to the same general balance of services as in a city is very misleading. This is particularly true if the external service providers insist that you should have a particular set of services, then say that they are too expensive to deliver and consequently threaten to shut you down. Surely that would not happen, you might think? Well, that is exactly the argument that has been playing out for Aboriginal settlements in recent years.

This argument emerges from misconceived notions of what makes a desert settlement 'viable'.

7.4 The 'viability debate'

Despite the array of great desert innovations and ideas, services that most of us take for granted – adequate education, good health, personal safety, job opportunities, even reliable rubbish collections and telephones, let alone discretionary services such as hairdressers, a choice of good food, counselling services and legal aid – are below any reasonable standard on many small settlements. In fact, they do not usually even exist on many remote Aboriginal settlements. Even though the aspirations of Aboriginal communities may be rather different from those of mainstream urban Australia, clearly their settlements are not meeting the baseline standards that all Australians expect to enjoy. Reports of child abuse, petrol sniffing, rape and personal injury, incarceration rates, low life expectancy, low self esteem and

BOX 16: NOT JUST MONEY! THE FIVE CAPITALS[168]

The western world tends to measure its success on the basis of financial capital. That is what the television news reports every night. Twenty years ago, though, practitioners in developing nations began to formalise the 'sustainable livelihoods' approach to recognise the fact that the people they were trying to help actually draw on many other types of resources.

The sustainable livelihoods approach is based on understanding five capitals – not only financial capital, but also human (skills, health, number of people, etc.), social (networks and organisations, etc.), physical (buildings, roads, etc.) and natural (the original source of most of our riches, through renewable uses of the environment such as cropping, as well as non-renewable mining). Sometimes cultural capital is talked of as a sixth type.

Poor people convert natural and social capital to help support themselves as much as they use money. We can show the relative importance by plotting them on a 'spider diagram', as shown here. The inner pentagon suggests the minimum acceptable level of capitals for survival, and the grey area describes a settlement that has little natural capital, but plenty of financial and human capital instead.

Of course, the five capitals are not just relevant to development – we all use them. Bringing the idea into our world highlights the value of volunteers and well-functioning institutions, which do not appear in the national accounts. One day, perhaps, the status of these will be shown on television as well.

Non-financial capitals are particularly important in remote areas. A pastoralist who draws on a social network to access feed in a drought receives as much financial benefit from keeping animals alive as from sales in a normal year. Some desert Aboriginal people spend large proportions of their limited cash income on travel to other communities to maintain their social and cultural capital, showing that they value these highly in financial terms. And the tourism industry is critically dependent on natural capital to attract most visitors to desert Australia.

Yet, nationally and locally, most of these non-financial capitals are not measured, so it is impossible to know whether our actions are increasing or destroying them.

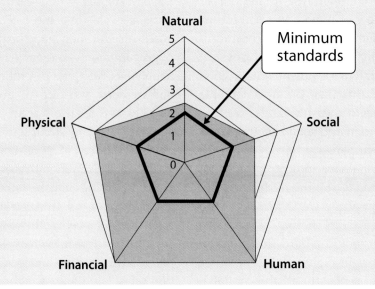

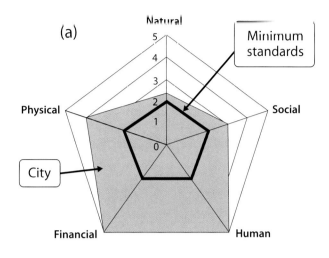

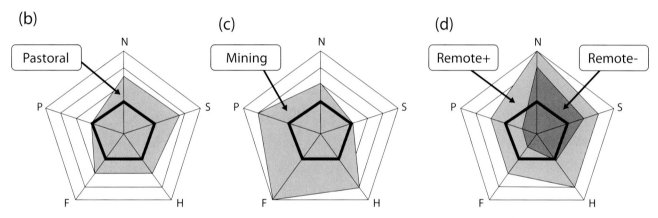

Figure 12: Using the sustainable livelihoods assets pentagon (see Box 16) to describe different types of desert communities and their needs: (a) the pentagon's five axes, a set of minimum standards (inner black pentagon), and a plot of a typical city (grey); (b) a healthy pastoral settlement (grey); (c) a typical mining settlement (grey); and (d) a functional (Remote+, light grey) and dysfunctional (Remote–, dark grey) remote Aboriginal settlement.[169]

catastrophic mental health abound – increasingly exposed by Aboriginal leaders themselves. As noted in Chapter 4, high-profile leaders such as Noel Pearson and Marcia Langton have publicly acknowledged that alcohol and drugs are destroying communities, and argue that communities need to take a response into their own hands, albeit with the support of wider society.[170]

Questions certainly need to be asked about why these issues are occurring, and why there are not the right services in these remote settlements to fix them. But the response of recent years has been insensitive – based on a simplistic accounting definition of whether a particular remote settlement is deemed 'viable'.

Some commentators have a simple mental model of desert settlements – that settlements need certain services, these services are more expensive to provide to settlements that are smaller and more remote and, hence, at some point certain settlements are deemed non-viable and should not be supported.[171] Services should be withdrawn and people 'encouraged' to move. This simplistic mental model is demonstrably wrong in several respects, and its unthinking application would also close down many other types of desert and rural settlement – whether based on pastoralism, tourism or services. Not all solutions to the woes of Aboriginal communities are to be found in service delivery models, but their analysis provides many insights.

If the idea of viability is useful at all, we need a more sophisticated model of what it means. Viability is a balance between the *aspirations* for services that a community may have, and the *cost* of obtaining them in a particular settlement. Essentially, the simplistic argument goes, 'if the costs of delivering the desired services exceed the funds available, then the settlement is not viable'. But, as we have seen, people can change their aspirations, trade them off against other values, and certainly they can differ from mainstream expectations. For example, most pastoralists are prepared to accept that it takes several hours to get to a hospital, that they only get their mail once a fortnight, and that the shops are a long way away – and that these are trade-offs for the lifestyle and enterprise opportunity enjoyed on a pastoral property. They are prepared to accept a lower quality of services in order to have a better life by other measures. Likewise, many Aboriginal communities are prepared to accept lesser services in exchange for being able to live near their traditional lands, and maintain aspects of their culture. Of course, it is legitimate for society to say that there are baseline standards of health and education and safety that every Australian citizen must receive, so this is not an excuse to accept violence or short life-expectancies. But it does say that the simplistic threshold model of viability is wrong.

On the other side of the equation, it is possible to vary the costs of obtaining services a great deal. Again, as we have seen, the pastoral community is expected to be quite self-reliant in obtaining many of its services – power, water, home schooling, rubbish disposal, and so on. This allows pastoralists to exercise a great deal of choice and self-reliance as to the services they wish to use and how they wish to access them. In principle, this should be true in remote Aboriginal settlements too. But policies affecting Aboriginal settlements in recent decades have undermined community self-reliance by imposing many government-supplied services based on standards established in distant cities. This is a supply- rather than demand-driven model of service provision.[172] Not only does it remove the

element of local choice and trade-off, but it is also usually more expensive as a consequence.

Remote community water supplies are a good example. The neat study outlined in Box 17 is a classic – it is easy to see how it happened. A well-meaning central administrator sees that the settlement has to have water. Expectations of equity say that this service should be as good as for any Australian. So government sets up a standard approach to delivering high-quality water, and promises to service it through regional service providers. The system fails, as is inevitable, and the cost and difficulty of getting it repaired from town means that the risk to the people becomes greater than it was in the absence of the service in the first place. As a consequence, other well-meaning administrators start to question whether it is safe and economic for people to live on the settlement in the first place. We find a problem, set up a solution, and the solution becomes the new problem! All along the way, people were well-intentioned, but acting out the wrong mental model.

It is not as if people who live in remote areas have not always worked out ways of getting the services they think are particularly important for themselves, given the chance or necessity. The next chapter looks at the some of the larger scale desert technologies that respond directly to desert conditions. There are also many smaller scale examples. For example, pastoralists were pioneers in the use of bores for water, for stock and themselves, building on earlier Aboriginal knowledge of soakages and springs. At Coward Springs in northern South Australia, early residents converted part of a natural hot spring into a hot pool that, in its own way, would rival the Roman experience at Bath in England. A local radio ham in Tennant Creek (clandestinely) set up a local television broadcast system in the early 1980s so that everyone in the township could tune into videos at a time when the options in terms of television stations were very few. A later version evolved (openly) into the highly successful Warlpiri Media, serving Aboriginal settlements in the Tanami Desert in their own languages. A wander around the junk yard next to the sheds at almost any pastoral station will turn up endless handmade curios to help shoe

BOX 17: MANAGING YOUR OWN WATER

Many small Aboriginal settlements in remote Australia suffer from unreliable water supplies and poor water quality; three-quarters of settlements depend on bores. Exploring these problems, desert researcher Robyn Grey-Gardner found that the water systems were set up by government and maintained by regional support agencies, which were often based in a service centre several hundred kilometres away.

'This conventional approach has been a disincentive to active involvement by residents in managing the hazards and risks of their own water supply', says Grey-Gardner. As a result, the hazards end up being worse – often the water supply will break down and, because a plumber has to come out from town, it can be weeks before it is fixed. And then the travel can literally cost thousands of dollars, just to fix a leak!

Grey-Gardner therefore ran a test project[173] in 2006 – with five settlements and the government agencies concerned – exploring whether more responsibility could be passed back to the community. She had to amend the approach taken in the Australian Drinking Water Guidelines to allow more of a risk-based assessment. But, once given the opportunity, she says, 'participants moved quickly beyond complaints about lack of service or not enough funding to working out what they could do themselves.' Once the roles and responsibilities had been negotiated with the agencies, and with a little training, they were soon carrying out many repairs and maintenance tasks in their own time and with their own money.

Of course, such self-reliance is nothing new – there were undoubtedly small pastoral homesteads just over the hill that were managing their own supplies in much the same way, and 30 years ago many of the Aboriginal settlements would have been too. The issue was that well-intentioned imposition of water services from far away simply failed to take account of desert realities, yet had left a legacy of disempowerment by moving decision making outside the community. This project is a beautiful illustration of how desert people can find their own solutions if only they are allowed to take back control over their lives, but sometimes it takes someone like Grey-Gardner to help negotiate the change with the outsiders.

horses, to run a radio, to clean tank water to make it fit for drinking, to turn a generator on remotely, to pull up the bore pump under a windmill, to make the rubbish disposal easier, to show the water level in the tank on the hill from a kilometre away, or to grade the potholes of a worn station track with a lump of old railway line behind the ute. All the things that we take for granted in a normal local government area, people have to do for themselves.

Yet, in an Aboriginal context, this self-reliance was taken away (and even pastoralists in some regions have a welfare mentality about subsidies), with services that were designed and delivered from central government. Imagine that schooling for pastoralists' kids in the outback had been set up on the urban model – collect them all by bus each day and take them to the central school, then return them home again after 3 p.m. Considering that stations are often 30–100 kilometres apart, it does not take long to see that this model is not only going to be prohibitively expensive, but that it is actually physically impossible to drive that distance within school hours. In reality, a different solution was found – the School of the Air (see Chapter 8).

This example sounds ridiculous when put so plainly, but there are many examples of urban-style services that have been repeatedly set up to fail in remote settlements over the past few decades: evaporative air-conditioners that fill up with scale in a month or two from salt-laden bore water and then need servicing by a town-based engineer several hundreds of kilometres away; complex power systems for which you have to call in an electrician when the fuse blows; buildings designed to need air conditioning when fuel and energy is hugely expensive in remote areas; fixed education infrastructure when the population is on the move; or patients' medical records kept in paper files in one remote clinic when the mobile patients could have their heart attacks or need emergency dialysis anywhere across three states. These are the sorts of supply-driven services, which are provided with good intent, that fail because the model is wrong, and are then used to suggest that remote settlements cannot be viable.

This is not to be seen as an argument against minimum standards of education or health – on Aboriginal settlements as on pastoral stations – but it *is* to say that real engagement of any community in charting its own future is critical to obtaining better outcomes. The water study described in Box 17 is instructive because it points the way to the solution: re-empowering people to find their own solutions to desert conditions, with suitable public support.

This requires a very different approach to determining 'need' than is normal today. It also opens the door for local communities to build their own houses, or do their own maintenance if they are allowed to tender for these services on an equal footing with outside providers. An equal footing in this circumstance involves valuing opportunities for local employment in the tendering assessment criteria, providing access to appropriate training and, of course, breaking the cycle of dependency.

The bottom line is that it is nonsense to suggest that the viability of a desert settlement is a simple threshold number that can somehow be determined from afar. Small desert settlements – each with a different community of livelihoods – will have very different aspirations from their urban counterparts, with no one size fitting all.

Of course, even if the costs of services are better managed by making the services genuinely appropriate to community aspirations and involving the community in their delivery, these costs must still be met from somewhere. Mining companies pay for many of the services to mining settlements.

Pastoralists make money to pay for their services. The financial economy of many Aboriginal communities is currently dominated by social payments. However, as noted earlier, some Aboriginal people earn income from selling arts and crafts, providing tourism experiences and certainly exchange social capital through their networks for other support. There are opportunities for a significant (but inadequate) number of jobs in providing the services that so woefully waste money at the moment – in house building and maintenance, management of essential services such as the water example above, and a range of local government activities.[174] Aboriginal people also substitute natural capital for some of their services in what researcher Jon Altman has called the 'customary economy', such as harvesting bush foods and using bush medicines instead of buying comparable goods at the store.[175] And there are a series of legitimate activities that look after natural and cultural heritage on behalf of all Australians, as explored in Chapter 5. In short, there are no simple silver bullets, but there are options that remote settlements have not to date been allowed to explore without interference.

It may well be legitimate to cap how much public support remote settlements should receive, but the judgment of what is 'viable' and whether people should leave those settlements has to be left up to those making the choice. To do that, policy must genuinely allow those people to work out what services they want and whether they can provide some of them more efficiently and effectively than distant service providers. Child abuse or the current level of personal violence or catastrophic health statistics are also not acceptable, but *making* Aboriginal people stay on or leave their settlements will not solve this.[176]

Underneath it all, society has a responsibility to ensure some basic rights are met because we believe this is appropriate for all Australian citizens. As many commentators have begun to argue, however, to create greater resilience in the long-term for Aboriginal people – rather than closing off even more options – requires an integrated approach. It requires building social, human, physical, financial and natural capital simultaneously – too often there is a

3-year program aimed at education, then a separate one aimed at infrastructure, without any effort to get everything working at the same time. There has to be a long-term inter-generational view, with concerted and persistent investment – turning around cultural change and education deficits cannot be achieved during the 3-year term of one federal government. Desert communities need greater responsibility and genuine rights to decide what services they need and how they should be delivered – and this requires support to develop the capability to express these aspirations sensibly. And an integrated approach needs to be taken to developing a set of livelihood options in different places: expecting a multifaceted solution, not a simple silver bullet. This will be helped by better linking the different types of communities – mining, pastoral, tourism, Aboriginal and any others – in ways suggested by the dependent and self-organising strategies of plants described in Chapter 3. Too often, at present, a single inspirational employment program fails because it never achieves the critical mass of change across the whole community that can only come from simultaneous persistent efforts on many fronts.

This case is acute for Aboriginal communities, but the same issues matter – if perhaps less life-threateningly – for all desert settlements. Service centres such as Alice Springs depend partly for their existence on helping to service the outlying settlements around them. If all remote settlements are closed down, the service centres will soon follow them into decline. This has occurred in western Queensland as the pastoral community has shrunk and globalisation has kicked in – small towns such as Windorah and Bedourie have shrunk away. These trends are hastened by the inevitable movement of people from remote settlements into larger towns, where there has been no preparation for them – in the absence of livelihood options, people are lost and social unrest ensues, as is predictably happening in Alice Springs today.

As we saw in Chapter 5, Australia needs people living out in remote areas, managing our environmental assets and providing safety nets and infrastructure, and potential employees for the mining,

pastoral and tourism industries. Hence, even from a narrow economic rationalist viewpoint, this remote population needs support. Indeed the billions of dollars of exports extracted from desert Australia, particularly through mining, mean that putting some funds back into the region is entirely legitimate to ensure that the exports can continue. If we also value the culture and settlement of desert people – Aboriginal and non-Aboriginal alike – then facilitating affordable access to demand-driven services (which may well be cheaper than supply driven ones) is essential. And, as we have already seen, there is no shortage of ideas among desert inhabitants!

7.5 Lessons from the desert

The two key messages from this chapter are, firstly, that desert settlements operate very differently from their urban counterparts and, secondly, that distant policy makers and service suppliers repeatedly misunderstand these differences, or are oblivious to them. Desert people themselves may not notice how differently their settlements operate, either because familiarity dulls this recognition, or because many people living in the desert come from urban environments and see the settlements through the same mistaken lenses as those outside. The future of desert Australia, and particularly of remote Aboriginal settlements, depends on both locals and those with an influence from a distance coming to appreciate how special these settlements are.

But there is another reason for outsiders to care. Profound changes in our world, driven principally by globalisation and demographic changes, mean that this understanding can also help decision makers with many settlements outside the desert. Two particular trends are at work. As we have already noted (Chapter 3), 2007 was the first year in which more than half the world's population lived in cities; in Australia, the proportion is closer to 80 per cent. City people have less and less daily experience of what is needed to make their food, supply their water and offset their carbon usage. It is hard to persuade people that they need to pay another dollar for their loaf of bread if they want farmers to look after the land better. The rural–urban divide is growing – everywhere. No longer do most people grow up in rural areas and have family members there who they visit regularly. Increasingly, rural settlements are operating differently from cities, yet city dwellers are less and less aware of the fact. And this matters because of the second trend: many non-urban settlements are taking on new forms and ways of functioning.

The little hamlet of South Durras, on the coast 16 kilometres north of Bateman's Bay in New South Wales, is an example. Over half its houses are owned as beach homes by people living elsewhere: mostly in Canberra. At weekends and in summer there is a massive influx of people; during winter there is only a skeleton population. Its small store struggles on the one hand to cope with the demand in summer, and then to survive through the lack of it in winter: how different from the steady business of the corner store in a city suburb. In Durras, the rubbish truck comes once a week – in summer it has to stop for full loads at every house; in winter most houses do not even have a bin out. Notwithstanding the 'sea-change' trend of people settling by the sea, all around the coasts of Australia there are settlements such as South Durras, where the patterns of seasonal occupancy create problems for local government – challenges in managing their rates base, variable needs for service provision, seasonal peaks in hospital admissions, progress association lobbyists that represent a few permanent residents but not, perhaps, the aspirations of most of the absentee owners.

These effects reflect mobility in the modern world, but between fixed settlements, rather like Aboriginal communities. As rising sea levels threaten some of our coastal towns, though, we may even have more to learn than we realise from mining

settlements that get uprooted and moved every few decades! How do these communities deal with the trauma of having their settlements destroyed under them? What infrastructure can be salvaged or designed for removal? What property rights systems minimise the legal overheads of changing locations? We do not have the answers to all these questions just yet, but learning from existing desert examples can help to conceptualise what might be needed in the rest of the country.

Land-locked rural communities are changing in other ways, but everywhere new patterns of settlement behaviour are emerging that do not correspond well with what happens in suburbia. Remote desert settlements represent a particularly extreme set of cases, but understanding the implications of their differences provides insights into what is needed for many other regions.

The solutions lie in three broad areas.

- There is the simple need to just be aware that rural and remote settlements may operate differently from urban norms (and from each other), and that an extra effort may be needed to understand community aspirations in these regions. This chapter has sought to raise such awareness.

- Some services for these settlements need technological solutions that differ from those in cities, often taking advantage of local characteristics; local people usually have plenty of ideas about this, if given the chance, and the next chapter explores a few of the solutions in the desert.

- New ways of governing such settlements and communities need to be found, and these are explored for the desert in Chapter 9.

8

Tantalising technologies

'A "healthy" or "good" country, is one in which all the elements do their work. They all nourish each other because there is no site, no position, from which the interest of one can be disengaged from the interests of others in the long term.'

ROSE 1996[177]

Innovations devised by, and for, desert communities may seem humble – sometimes even rudimentary alongside the powerhouse of modern industrial high tech, the internet, power generation, space flight, bio- or nanotechnology – but they do aim to have one outstanding quality: sustainability.

Just as the burning of fossil fuels has led to climate change, many modern technologies are greedy in their demand for, or impact on, natural resources. Sooner or later, they confront society with undesirable consequences and side-effects, such as shortages or pollution, which scientists and policymakers then have to deal with.

In deserts, things that are greedy for scarce resources do not survive for long, and desert technologies, as well as peoples, have to obey this law. The qualities that define deserts – variability, scarce resources, sparse populations, remoteness, political and technological isolation, and the importance of local knowledge – demand technologies that are fitted to purpose. They need to be imbued with a philosophy of minimalism: of taking only that which is necessary to achieve the goal.

The previous chapter described how misconceptions about the desert undermine desert ingenuity, so this chapter is dedicated to examples of its success. Like other citizens, desert dwellers require basic services such as education and health care. They need technical services such as water, power, buildings, transport and communications to support these. It is the way desert dwellers have solved – and are solving – these problems that is instructive and which contains principles applicable to the wider society.

The vital backbone of communications underlies services to sparse and mobile populations. Some resources are plentiful in the desert, such as solar energy and space; these must be capitalised upon. Others are limited, such as water; these must be conserved. And social services such as health and education must themselves become mobile to deal with a sparse and mobile population.

8.1 Communication technologies

For most of the time since the dawn of human experience we have been nomads – settlement, and an urban life, arrived only in the most recent minutes of our story. Nomads have always put a high value on exchanging information and maintaining social capital. We have already seen how Aboriginal oral traditions kept information and experiences alive over thousands of years, probably more effectively than books and tablets in the Middle East. Without a written language or the benefit of email, Aboriginal people had a simple, but precious, message stick technology when important messages had to cross their lands. Elders would carve symbols indicating meeting places and dates, and a messenger would carry it hundreds of kilometres to other tribes to set up ceremonies or transmit critical news.

When Europeans first entered central Australia, news did not travel much faster than this. But, in 1872, the Overland Telegraph Line from Adelaide to Darwin was connected. The delay in getting messages from Melbourne to London suddenly changed from 2 to 3 months by ship to a little over 24 hours up the line of repeater stations through central Australia to Darwin then Java, Malaya, India, and so on, through to Europe – nearly 100 times faster. This was not a desert-specific technology, of course, but it was the start of connecting inland Australia to the world. When the station master at Barrow Creek Telegraph Station was speared on 23 February 1874 by Aboriginal warriors angry at interference with their land and their women, the dying man was able to send a message to his wife in Adelaide in minutes, instead of weeks.[178] Within a few decades, people were adapting their use of other new technologies to desert conditions.

Alf Traeger developed the robust pedal radio in South Australia in 1927,[179] turning the temperamental technology that had blossomed during World War One into the necessary and reliable communications network that, for three-quarters of a century, formed the central nervous system of outback communication. The advent of pedal radio voice communication in the 1930s brought a real change to the quality of life in isolated places. It supported medical services and education as discussed below, but it also reduced the loneliness of desert life by enabling people living hundreds of kilometres apart to speak together. Help and friendship were now only a radio call away.

Women took on the role of radio operators on the big pastoral stations and quickly developed a bush community over the air. In time, the 'Galah Session' developed, which allowed a chat among neighbours who could be hundreds of kilometres apart. The Royal Flying Doctor Service's superintendent, the Very Reverend Fred McKay, once said of Traeger, 'He created a social revolution. Human relations were transformed. In a very real way he made Outback Australia'.

Today, another generation of cutting edge technology is being pressed into service to open new opportunities for desert living.

After many decades faithful service, Traeger's radio network gave way to phones using satellite and microwave concentrators in the 1980s and 1990s. But these technologies were initially expensive for people outside the larger settlements, and people on some Aboriginal lands found themselves reverting to radio for flexible (and free) communications out bush. Today, researchers such as Dr Mehran Abolhasan of the University of Wollongong are exploring how a 'kangaroo hop' mobile phone could bring a new mantle of safety to the outback: allowing people to communicate or do business across the vast distances using a mobile phone that does not need a comprehensive network of expensive fixed towers or satellites. In keeping with the desert strategy of nomadism, the technology exploits what are known as *multi-hop ad-hoc networks* in which each mobile phone handset serves as a carrier in an ever-shifting network that has only a few fixed transmission points. As Abolhasan explains, clever software manages the calls through this dynamic network and picks the best route – or set of hops – without phone users even being aware of it. All they have to do is leave their mobile handsets switched on.

The multi-hop network is just one way that research is exploring how to deliver medical alerts, emergency services, remote monitoring, text messages and financial services over huge distances where normal infrastructure is scarce or non-existent. To overcome cultural differences in a mobile population, a novel messaging system called MARVIN allows remote communities to broadcast voice or images to one another in local languages through the medium of television or multi-media phone messages. Using MARVIN, locals with minimal training can quickly make computerised animations of local people or characters and have them speak realistically in the local language. Originally developed in outback Australia to help to communicate health messages by Jay Easterby-Wood, the smart animations are now being used all the way from nursing on the Pitjantjatjara lands to education in Britain and supporting the Quitline advertising to stop smoking.[180]

These technologies both aim to cope efficiently with sparse and mobile populations over very large areas, and allow them to stay in touch with the networks that sustain their reserves of social capital.

Communication technologies hold the key to the future of desert communities in Australia and elsewhere. With isolation and remoteness as the dominant characteristics of desert life, the inhabitants are for the most part still ill-served in terms of communication access, compared with their urban counterparts. As a result, governments have tremendous difficulty in delivering effective services to desert people, who, in turn, remain isolated from markets and the global economy. The web, high-speed broadband and the digital age are changing all that, and altering forever the character of life in the deserts. Efficient uses of technologies such as video conferencing and the internet are other critical paths to the future, as we have seen in Chapter 6 with Mick's Whips, internet trading, Desart and the cross-border business clusters.

'Broadband is the great hope for Aboriginal communities and others who want to remain in remote areas and to preserve their culture, for the reason that it is very amenable to all forms of human expression', says Dr Bruce Walker, CEO of the Centre for Appropriate Technology in Alice Springs. 'It enables you to transmit word and song, dance, art, music, knowledge of many kinds, and even sport, over immense distances – between communities dwelling in a desert, and between desert communities and the whole world.'

Increasingly, services that once required communities to live in a location deemed convenient by centralised governments can be delivered digitally. With the living examples of Australia's Royal Flying Doctor Service and School of the Air to follow, health-care advice and education – along with technical information about water, energy, waste disposal or food production – are increasingly arriving via broadband to even the most isolated of dwelling places.

At the same time, desert communities have an unprecedented opportunity to share their culture, creations, technologies, ideas and produce with the world: to become linked in real time into the global marketplace, Walker says. The challenge is for those communities to project their own identity out to the rest of the world confidently before they are swamped by a homogenised global culture pressing back in.

An example of the way the internet is globalising Aboriginal culture and transporting it to far-away places confidently is the case of the Chooky Dancers: a Yolngu dance troupe from remote Elcho Island who put their highly idiosyncratic version of Zorba the Greek on YouTube. This attracted a global audience of well over a million people, along with thunderous recognition. Comments ranged from 'So You Think You Can Dance - eat your heart out!' and 'The most original performance of Greek dance I have ever seen!' to 'They're awesome!'[181] As a result, the Chooky Dancers performed in Canberra at the Stolen Generations Apology, and then headed overseas. This is novel outback culture on the march, evolving as it goes!

8.2 Energy technologies

Energy for living has always been an issue in desert regions – to a dispassionate alien, that might seem like a paradox considering the abundance of sunlight available. Nonetheless, midday in summer is fiercely hot, while midnight in winter can be bitterly cold, so humans have to find ways of heating and cooling themselves and their food. Early desert nomads, such as Aboriginal people in Australia centuries ago, avoided working in the hottest times, and guarded fire sticks for cooking and warmth in the colder times. Where people settled in deserts around the world, they developed passive building designs that stayed cool and sucked in breezes in summer, and stored heat in their thermal mass to stay warm in winter. However, the advent of building designs, water heaters, air conditioners and cooking appliances foisted on deserts by the populous industrial centres required greatly increased energy supplies. Perversely, early buildings in desert Australia were often more climate- and energy-friendly than today's air conditioned boxes.

In Australia, and elsewhere in the world, this recognition spawned the appropriate technology movement in the 1960s, which was aimed at developing approaches that are small scale, low cost and technically matched with local skills. In Alice Springs, the Centre for Appropriate Technology (CAT) was established to explore how such simpler technologies could meet people's needs in remote Australia. With links to what was happening internationally, CAT has developed and manufactured solar ovens, simple hand water pumps and efficient heaters in central Australia over the past 20 years.

As with communications, however, today's smart energy solutions deploy a new generation of high technology in novel ways to meet desert needs. A key example is the Bushlight project recently developed by CAT in central Australia.

The technicians at CAT used efficient solar panel and battery technologies to deliver a complete power system suited to remote conditions – Bushlight's Remote Area Power Supply system (RAPS) for reliably supplying the needs of a small settlement that has no access to town water or power. Even just to refrigerate food in these settlements requires a power supply that is sturdy, simple to use and dependable. The challenge was not so much in the solar technology, but more in creating a reliable system – a unit that can be hoisted on to a truck, driven hundreds of kilometres across rough bush tracks, easily unloaded and connected

up at the other end, and which will then run reliably for years in places where repairmen rarely venture. The Bushlight project has installed over 90 units across the outback in the Kimberley, central Australia and Cape York – in the process bringing down the cost per unit significantly (from $40 per watt installed to $26), lowering greenhouse gas emissions, saving households an average of $5500 per year in energy costs, and providing many ancillary community benefits.[182]

Bushlight is not merely a technology fix. It comes with a complete package to assess the needs of the community and help to design the right unit, to train locals in its use and maintenance, install it, and then provide expert support for it. The systems use intuitive controls that enable users to manage their own electrical energy use effectively. They also possess neat features to help control demand, such as circuit timers and energy management units to ensure energy is shared equitably within the settlement. The systems come in three configurations – a solar system aimed at a single household, a solar-only system for a settlement of two to ten buildings, or a hybrid system with solar and integrated generators for a settlement.

Bushlight meets many of the criteria for successful desert survival: it is highly mobile and it depends on a resource – sunlight – that is ubiquitous and in plentiful supply, though at a low density compared with other sorts of energy. It is highly suitable for small or very small communities, it uses local knowledge to operate at its best, and its performance is independent of the policy back flips of distant governments.

In short, Bushlight is the sort of integrated technology solution capable of operating anywhere in the world where access to the main power grid is either too costly or simply impossible. Such approaches could help meet the needs of a significant proportion of the world's 3 billion inhabitants who have no electrical power as yet. As the costs of solar come down, it may also become a viable solution for other rural dwellers who wish to cut their greenhouse footprint and escape the growing costs of electric power based on fossil fuels.

Although Bushlight is aimed at smaller settlements, towns such as Alice Springs also face energy problems. At present, Alice Springs obtains most of its power relatively cheaply and efficiently by burning gas from the nearby Palm Valley gas field, which also supplies Darwin through a pipeline that runs the length of the Northern Territory. However, this gas field's life is limited, and, like the rest of central Australia, Alice Springs experiences the paradox of needing energy to cool itself in the face of too much energy from the sun! Surely the sun should be the major source of energy rather than a reason to burn non-renewable fossil fuels.

Encouragingly, Alice Springs has embarked on the first steps towards becoming a solar city, as the first remote town to take part in the Australian Government's Solar Cities program. The 6-year project aims to establish the market conditions and sources of information that encourage citizens to adopt solar or renewable technologies, and to encourage a community-wide discussion that leads to desirable changes in behaviour. The project involves installing around 225 residential commercial solar systems, several solar installations in key public buildings, 1000 domestic solar hot water systems, 850 energy audits and up to 400 smart meters in businesses and homes around the city to help residents monitor and adjust their power use.[183] In related activities, the town has refitted the Alice Springs Civic Centre as a five-star-energy-rated commercial building, established a solar collector demonstration display, and created a Smart Living Centre. An old Territorian house is also being monitored and gradually retrofitted with solar and water-saving technologies by locals, to work out what people can do in existing homes. For example, new reflective paints can reduce roof temperatures by ten degrees or more; one of the original commercial versions of these paints, 'Solacoat', was invented in Alice Springs in 1988, although it is now sold around the world.[184] There is a long way to go to reverse the effects of decades of imposed externally driven architecture and energy supplies on the town, but this is a significant start.

The application of new energy technologies in smart ways in desert environments (Plate 8a) can make remote dwellers less dependent on energy and servicing from elsewhere, both by harvesting local energy – the sun – and by using more local knowledge to do maintenance on robust energy supply systems such as the Bushlight RAPS. Even in larger settlements such as Alice Springs, local knowledge and experience is being galvanised to find solutions that use new technologies cleverly.

8.3 Water technologies

Clever ways of conserving water have been a part of desert culture since history began. Many of these ancient approaches and technologies will find fresh application in the critically water-short years later this century when, it is estimated, four humans out of every five will confront moderate to severe water scarcity. The ancient covered canal systems, or qanats, of the Middle East and central Asia have transported water with minimal evaporative losses for thousands of years, yet the open irrigation ditches of the past century in the Murray–Darling Basin lose millions of litres a year – a case of foolish forgetfulness of desert knowledge.

While the average person needs only 2.5 to 5 litres of water per day for drinking purposes, they indirectly consume around 3000 litres of water per day, which is used to grow their food – roughly one litre of water for each calorie of food energy produced.[185] This poses a fundamental challenge in a world where more and more water is being directed to non-food uses and where food producers cannot afford to compete for high-priced water with industrial and city users. Desert people were once naturally conservative in their capture and use of water for food production, drinking and household purposes; new lessons could be learned from old methods when designing more water-conservative systems for the wider society, as shortages begin to bite.

For food production in the form of outback beef, modern high technology is again creating new options for the efficient use of precious water, as Australian graziers move towards precision pastoralism (see Chapter 5). In arid Australia, water should be like gold and, across the continent, a new generation of misers are taking pride in finding novel ways to hoard and make the most sparing use of it. A great deal of the water used for stock in pastoral Australia is captured from run-off into small stock dams. Dr Ian Craig at the University of Southern Queensland has estimated that farm dams store about 7000 gigalitres of water in Australia, yet lose as much as 40 per cent of this to evaporation every year. Even if they are only full for half the time, the total loss would be 1400 gigalitres a year – nearly three Sydney Harbours, his study says. So, new technologies that actually prevent water loss are proving to be invaluable as they are tested in practical conditions by pastoralists. These include floating blankets of 'bubble wrap', plates of polystyrene, shade cloth, and thin layers of novel chemical monolayers that cover the water surface and slow evaporation.[186]

The wider approach to precision pastoralism is exploring technologies to help improve animal management, control grazing pressure, cut costs and enhance the environmental condition of pastoral country. 'The way water is delivered to stock, where watering points are located and how they are managed has a big impact on the business of running a station', says Colleen James – project officer for this WaterSmart project. 'It affects the condition of the country, the sustainability of the enterprise and its running costs.'

Pastoralists have quickly found that not only do the new technologies save water, they can also save 30–50 per cent of the costs of managing water in a remote location, thus offering a dramatic improvement in economic productivity and more options for sustainable grazing. As noted in Chapter 5, this can lead to a whole change in culture.

If a windmill or bore gets blocked and is not fixed quickly, livestock die. So pastoralists had to use up fuel driving their 'bore run' regularly – as much as every day in summer. At other times, artesian bores ran free, spilling millions of litres of precious underground water out on to the desert soils. Throughout the history of pastoralism, the regular visiting and maintenance of watering points has been a long, expensive and unavoidable chore.

However, the introduction of telemetry and remote surveillance has brought dramatic change. Bores are now monitored by computer from station homesteads. This can provide instant feedback and imagery on dam and tank levels, water flows, pump control, medication of stock, local rainfall, stock usage patterns and even water quality. These techniques allow a fine-tuning of grazing pressure around the huge expanse of the pastoral property that can, in the hands of a good manager, result in higher quality animals, water savings and improved native vegetation condition.

'Telemetry gives you peace of mind. You also spend less time on the road. We had a problem with one of our tanks and the telemetry sent an alarm. Being able to respond quickly to situations like that makes a big difference', says Kylie Fuller of De Rose Hill Station in the Northern Territory. On Monkira Station in Queensland, a study showed remote monitoring has cut the time spent driving round to watering points from 40 hours a month to less than 20. On Mount Ive Station, Len and Joy Newton say their new telemetry system paid for itself within 8 months, and running costs fell by almost three-quarters. As described in Chapter 5, pastoralist Roy Chisholm was able to monitor and control his bores on Napperby Station in central Australia using the internet while on holiday in South America.

As with all new technologies, pastoralists are discovering the drawbacks too – such as cockatoos that developed a taste for chewing up solar panels, cables and aerials. With classical outback inventiveness, Ray Jansen at Canobie Station solved this problem by hanging rubber snakes around the vulnerable electronic gear – the birds left it alone after that.

Water telemetry is just part of the battery of high technologies that is changing the face of grazing and potentially making food production from desert regions more sustainable. Solar-powered pumps, automatic stock weighing machines, airborne or satellite sensors and, potentially, virtual fence lines created by electronic implants in cattle ear tags are all part of the technological arsenal of precision pastoralism. Andy Bubb, project leader for the precision pastoralism work of the Desert

Knowledge Cooperative Research Centre, says that the ability to know the condition of individual animals, and to use automatic drafting systems to separate them off into a holding paddock for sale, is 'a fundamental shift in the way we think about livestock management in remote and extensive grazing situations', which may one day have major export potential.[187]

The new water technologies obey many of the precepts of desert living: they enable livestock management to be more nomadic and graze a wider area; they provide reliable water; through remote control, they can protect landscape refuges; and through dam-liners and blankets, they conserve their reserves for times of need. Most importantly of all, they make life easier for the handful of individuals who have responsibility for the stewardship of half the continent.

Outside the pastoral context, water managers are using modern scientific understanding to reformulate other ways of saving and protecting the quality of limited stores of desert water. The ancient civilisations of the Middle East understood the importance of saving water in large underground cisterns to avoid losses from evaporation, or contamination by livestock and human waste. Mirroring this approach is the twenty-first century Australian technology of 'managed aquifer recharge', where water that would otherwise run to waste is injected underground into a convenient, but well-sealed, natural aquifer, where it remains until needed. Filtration through the sediments also cleans the water so that modern aquifer recharge prevents both evaporation and contamination – just like the ancient cisterns.

This technology is particularly suitable for desert and water-scarce regions. Just south of Alice Springs, Australia's first arid zone managed aquifer recharge – intended to purify treated sewage to a state where it can be re-used to water crops – went operational in 2008. The concept arose from a need to prevent overflows from the Alice Springs sewage stabilisation ponds from contaminating the nearby Ilparpa Swamp. Up to 600 megalitres of water a year is now stored in the underground aquifer for subsequent use in irrigating local horticultural crops.

Such systems for storing water obey the principle of local harvesting and re-use of scarce desert resources. Because big towns cannot move to follow water resources across the desert, they must make better use of the ones they have. Aquifer recharge offers such a possibility in Australian towns and cities, and in arid communities worldwide. In desert regions subject to large but very sporadic downpours, the harvesting of stormwater from urban streets and drains and its injection into aquifers offers a further opportunity to conserve this scarce resource.

8.4 Social services technologies
As we saw in Chapter 7, desert settlements require not only physical services such as water and energy, but also social services such as health and education. Over history, these have often posed a problem for desert people around the world. As central governments have sought to deliver better services to nomadic people such as the Tuareg in Niger and Mali, they have urged them to settle: sometimes voluntarily, sometimes forcibly. Whatever their political motivations, central governments often genuinely want to improve the health or education of a mobile population, so, from a central perspective, this seems sensible and in the best interests of the nomads. But, of course, settlement immediately destroys the nomads' way of life, so they often resist the idea. Australian governments have faced the same dilemma with Aboriginal people, and even with remote settlements of pastoralists and miners, although the latter may be less nomadic. But there have been some remarkable innovations to overcome these problems, on which modern technology can build further.

Australia's most famous desert services innovation – attracting admiration and emulators around the world – is the Royal Flying Doctor Service (RFDS).[188] Founded by the Rev John Flynn in 1928 to provide his 'mantle of safety' across the outback, it was an insightful response to the desert challenges of sparse settlements, isolation and lack of hospitals (Plate 7c). The RFDS thrives to this day because it has continually evolved – responding to the changing patterns of settlement, as well as capitalising on new technologies and demographic trends, such as the increase in outback tourism. It became far more than merely a plane to pick up sick people: developing all the other elements of efficient communication, connections to health services, protocols for airstrips, and design of equipment so one or two people could handle patients without assistance.

The RFDS arose from the Australian Inland Mission that had established nursing homes and a roving team of ministers to attend to the scattered population's physical and spiritual needs. Initially, the Mission was helped by labelled medical chests and body charts that enabled some diagnosis and medication to be provided remotely. Today, the RFDS continues to be at the cutting edge of communications technology. Telephones, satellites and the internet are increasingly replacing radio communication, and there is a decline in the use of the Flying Doctor radio service. Whereas in the 1980s all calls for medical assistance came in by radio, by 2008 only 2 per cent were radio calls.

The RFDS claims with justifiable pride that, when an emergency call is received by a Flying Doctor Communications Officer, the caller can be in contact with a doctor, nurse and pilot within 30 seconds and an aircraft can be airborne in 45 minutes or less. Although over 100 000 Australians still live out of reach of doctors, thanks to the RFDS network of bases across Australia, no-one is more than 2 hours away from medical help.

Outback conditions and the RFDS spawned another great service idea – the famous School of the Air – a virtual school that eventually covered an area the size of western Europe. It started in 1951 in Alice Springs with the aim of overcoming the disadvantage in access to educational services that all desert dwellers face as a result of isolation. Until recent years, the School used the RFDS radio network to link children and their teachers and deliver a program of education that includes all the usual subjects taught in city primary schools. With improvements in technology, there is now no need to use the radio – telephones and the internet have become the preferred means of communication.

The organisation was re-named in the 1990s as the School of Distance Education, but somehow the more aspirational-sounding 'School of the Air' has persisted.[189]

In a further development aimed at delivering education to out-of-the-way places, Charles Darwin University and the Centre for Appropriate Technology have developed the idea of Mobile Adult Learning Units or MALUs (*malu* is, appropriately, the Luritja word for the highly mobile red kangaroo). These respond to the reality that people on remote settlements find it hard to get into town to undertake technical training in essential services such as water, power and construction for their communities. The MALUs are fully self-contained training units on trailers that attach to a prime mover so that they can be towed out to a settlement for several weeks at a time (Plate 8b). They provide living quarters and classrooms for two staff members to deliver on-the-spot practical training in areas of need identified by the community.[190]

Expanding on the principle of mobility and communications in health care and education is the advent of telemedicine. Consider the simple issue of sharing medical records between different jurisdictions. Because of their mobility, Aboriginal inhabitants of the Ngaanyatjarra-Pitjantjatjara-Yankunytjatjara lands around the South Australian–Western Australian–Northern Territory border can easily be in three different states in a single week, or indeed live in any of them for extended periods. If someone in this region needs emergency care, they are just as likely to be evacuated to Kalgoorlie (or Perth) in Western Australia, as Alice Springs in the Northern Territory or to one of the Adelaide hospitals in South Australia. As a result, their medical records are almost certainly going to be in the wrong place. Until recently it could take days to get these transferred, by which time the insights that medical history can provide in a case of an acute illness may be too late. Increasingly, these records can be held online in consistent systems in the different states.

As telemedicine grows, remote diagnosis – an outback patient consulting a specialist in Sydney or Darwin without leaving home – becomes

BOX 10: NGAANYATJARRA-PITJANIJAIJARA-YANKUNYTJATJARA WOMEN'S COUNCIL

The Ngaanyatjarra-Pitjantjatjara-Yankunytjatjara Women's Council, more familiarly known as NPY Women's Council or NPYWC, is a unique desert organisation delivering social services seen as important by locals.[191] Established in 1980 by Aboriginal women to meet their needs, and those of their families, it began working across the state borders in a remote area south-west of Alice Springs (see Figure 13) at a time when health, education, justice and all social support services were independently, distantly and quite differently driven from Perth, Adelaide and Darwin – depending on which side of the border one happened to be on that day.

The NPYWC has run a respite care service for the aged and infirm, support for youth, nutrition awareness, domestic violence support service, challenged substance abuse and brought NPY women together to build their common strength for more than two decades. Their community-driven efforts are continually made harder by having to knit together services based on the disjointed funding guidelines and programs of three different jurisdictions (their website lists at least 35 funding sources for their $5m budget!), but, despite this, they have a long list of achievements.

The NPYWC is a bottom-up service-delivery organisation that has persisted despite (rather than because of) the system, yet which has achieved efficiencies of service delivery at an appropriate geographic scale across a set of services that would not sit comfortably in a single formal government structure. In the cross-cultural context, it has been sensitive to the differentiation of roles of men and women in Aboriginal society, and to the mobility of people in the region. It struggles with the fragmented policy environment in which it has to seek funding, but it was instrumental in promoting the innovative cross-border policing arrangement that the three jurisdictions have established in this vast, remote region.

increasingly practical. This uses internet and other forms of communications to transmit diagnostic information, interview the patient 'live' and send local test results and scans back and forth. Eventually, basic surgical operations could be carried out remotely by a surgeon in Sydney directing a scalpel that is physically located on a machine over a patient lying in the medical centre of a small outback town. Such developments, however, will require the construction of very large fibre optic communications channels to permit the transfer of the vast volumes of data necessary to control the operation in real time across thousands of kilometres in nanoseconds. Such systems have already allowed critical care to be delivered virtually in Katoomba rural hospital by doctors in Nepean Hospital in Sydney.[192] In practice, the very high speed channels required will remain restricted to larger settlements for years to come.

Coupled with this is the proposed 'hospital without walls': the real-time care and monitoring of a non-critical patient in their own remotely located home. Using automatic patient monitors, doctors and nurses can maintain surveillance over distant patients, ensure proper treatment is being carried out and identify changes in condition that require a different treatment.

Working in the other direction, doctors servicing Utopia have worked out a unique solution to the stresses of administering to a remote population. For two decades, a kind of migratory cooperative has been in operation, where individuals come up to these remote Aboriginal lands north-east of Alice Springs for 3 to 6 month terms. For the rest of their lives, most doctors contribute to practices in peri-urban Victoria. Many other regions have seen a doctor spend several years in the community and then burn out with the rigours of being on call 24 hours a day, 7 days a week; such experiences usually mean people leave and never return. Instead, at Utopia, doctors do a little at a time, but return repeatedly; they come to know, and be trusted by, the community, yet do not burn out. Gradually, new members join the rotating team as others may retire or move on. The result is a far better continuity of doctoring, and another approach to mobility in remote regions.

Such developments open up for desert people the prospect of much easier access to general and specialist care that once meant a long and costly journey away from home for the patient, and often their family as well. In theory at least, the telemedicine infrastructure being developed by researchers in Australia for use across the outback could in time extend as a major export to remote desert regions almost anywhere on Earth, where communication infrastructure permits. One day, perhaps, most desert people the world over will enjoy access to a standard of care equivalent to that experienced by those who live in cities. Importantly, too, for those whose livelihood or personal attachments pressure them to stay at home, such developments may spell the end of the sad exile of patients suffering from serious or terminal diseases. Today, treatment of such diseases necessitates leaving for a distant hospital far from friends and family, where many patients may die without ever returning home.

The Flying Doctor, the pedal radio and the School of the Air were desert solutions to desert problems – the issue of a sparse and mobile population supported by scarce resources that could not pay to have the services everywhere, so they made the services mobile or remote too. Box 18 illustrates a community-based organisation that also been innovative about how it organised itself to deliver a wide array of social services into the remotest part of desert Australia.

8.5 Smart ideas for a dry world

In the twenty-first century, all humans will face the reality of limited resources. Four out of five will face water scarcity. Everyone will have to deal with costlier energy and food. Some people will be more mobile. And most will find themselves living in a warmer, drier and less predictable world.

Like desert plants and desert people, desert uses of technology have evolved to cope with these uncertainties, as this chapter illustrates. Most of these technologies are not earth shattering in terms of novelty, but almost all of them, in one way or another, are Earth-conserving, in that they help people to make better use of sparse resources and to deal with sudden change.

In dry times, such approaches to technology could be Australia's gift to the world. Many other nations and cultures have wonderful desert technologies – the dwellings of the Pueblo Indians or the micro irrigation developed by the Israelis, for example – and Australian outback technologies form but a part of this global desert heritage. Like those of other deserts, they have been hammered out on the anvil of human experience, tested under harsh and unforgiving conditions, and found to deliver practical outcomes. Like many Australian technologies, they combine high science with robust, low technology that keeps on working.

Desert technologies help to eke out scarce resources in many ways: sometimes by using them more sparingly; and often by assisting with human mobility so that too great a pressure is not placed on the resources of one place by permanent exploitation. They tune the awareness of the user to a world in which supply is erratic and unpredictable – a world that the modern urbanite has long left behind, but may soon be about to rediscover.

Even the mere act of thinking about how to deliver services to isolated Aboriginal communities – hundreds of kilometres from the nearest sealed road, telephone or water treatment plant – generates ideas that benefit the whole of society. Mobile health records, for example, are relevant not only to nomadic desert people but also to a generation of retirees who have become grey nomads travelling the continent. The learning of the isolated community that manages its own energy, water and waste may be a blessing for 'sea changers', hobby farmers, retirees and people with alternative lifestyles who want to escape the bondage of soaring power and water bills from big utilities.

We have seen that some resources, such as solar energy, are plentiful, and must be taken advantage of. Others, such as water, are scarce, and must be conserved. People are spread sparsely, making conventional fixed services both expensive and illogical; people are also mobile, so the services themselves need to express mobility and be tailored to local conditions. Underlying all this is smart use of the enabling technologies of modern communications (Plate 8).

To capitalise on the potential of the natural inventiveness of its desert people, Australia needs to invest in, and apply, the latest communications technologies to offset sharply rising travel and transport costs, improve service delivery and reduce the human impact of isolation. Communications create opportunities for isolated communities to deliver their services, technologies, creative ideas and desert products around the world.

We need to support innovation with technologies (such as solar power) that seem marginal in cities, using the desert as a particularly rugged and commercially viable proving ground for ideas that

may ultimately achieve widespread adoption in a resource-limited and uncertain world. Some of these opportunities lie with more sustainable, low-cost energy, water and waste disposal systems for small, and even large, communities, for a more resilient world – and we must share the knowledge. At the opposite end to the 'small is beautiful' philosophy, some opportunities could even build a contribution of the Australian desert to the national and global economy at the very largest scale (see Box 19).

Such developments will be supported by stronger, larger networks that share knowledge and skills locally, outback-wide and through deserts worldwide. This will create a global market and brand for Australian expertise and technology that caters for mobile populations, isolated communities, moving settlements and harsh, unpredictable conditions.

Support for innovations that respond to the special conditions of the desert will create a significant opportunity to benefit the world and Australia's economy at the same time.

BOX 19: AN ENERGY-BASED FUTURE FOR DESERT AUSTRALIA?[193]

The Australian desert receives more sunlight than practically anywhere else on Earth; in principle, an area of desert just 50 kilometres square could meet all of Australia's energy needs. Leading global engineering firm WorleyParsons has proposed building the biggest solar thermal plant in the world – a massive 250 megawatts – to take advantage of this abundant resource. The plant would be the first of 34 similar ventures to be built by 2020 and designed to supply 5 per cent of the nation's total energy, and half its planned renewable energy. Its proponents argue it is feasible because it is based on mature technology that is 'not high tech', with readily available and recyclable components, and would pay for itself in just 5 months. They say that peak load on the power grid coincides closely with the times of maximum solar energy collection; and at night energy can be stored in molten salt baths.

Another major desert resource is geothermal energy, which can potentially be extracted from hot granite rocks 3–5 kilometres below the desert surface. More than 30 companies have joined the hunt for these resources. The Australian Geothermal Energy Group estimates at least 2250 megawatts of geothermal capacity could be on line in Australia by 2020, producing 17.5 terawatts of electricity by 2020, or 5 per cent of national consumption. Geothermal power has the advantage of being available 24 hours a day, so perfectly complementing the fluctuating supply from solar collectors and reducing the need for energy storage.

What makes remote area power increasingly feasible and attractive, says physicist Barrie Pittock, is the development of low-loss high voltage DC (HVDC) transmission lines, such as the Basslink connector that sends power across Bass Strait between Tasmania and Victoria. A little imagination brings further benefits. These lines lose less than 3 per cent of their energy for every thousand kilometres or cable. Over time, Pittock believes, 'This means that besides supplying Sydney, Melbourne and other centres from the deserts, we can also export energy direct to Asia using submarine cables. The distance from Darwin to the Indonesian grid is relatively short.' And if the power was being carried across northern Australia, it could be used there to provide clean power input to aluminium smelting in the north, instead of the current practice of shipping huge quantities of ore across to New Zealand for smelting with hydropower. Besides powering industry, these new generators could create livelihoods for desert inhabitants.

Clean power company DESERTEC envisions weaving together large-scale solar, geothermal, wind and wave energy projects across the centre and north of the continent. These would form an interconnected lattice capable of satisfying the nation's entire energy needs. 'With a coordinated rollout of renewable energy capacity and long-distance transmission infrastructure Australia could – sometime after 2020 – be producing surplus, premium-priced, low greenhouse gas emission power', says DESERTEC's Stewart Taggart. 'This surplus power could be sold into Asian markets through sub-sea HVDC power lines connecting Australia to the region.' Combined with North West Shelf natural gas pipelines, by mid-century he envisions a 'pan-Asian energy superhighway' stretching from the Australian deserts as far as Beijing.

Such proposals are now being treated seriously around the world: the European Commission's Institute for Energy is considering harvesting most of Europe's energy needs from solar farms located in the Sahara desert, and transmitted across the Mediterranean in HVDC cables. These would be combined with geothermal energy imported from Iceland and wind farms fringing the North Atlantic to even out the energy flows. EC scientists estimate that such a system could produce 100 gigawatts of solar energy by 2050 for a total cost of 450 billion euros. Based on current costings, this would produce energy somewhat more cheaply than present fossil-fuel based systems.

Pittock argues that Australia could easily meet its target of 20 per cent of all energy drawn from renewable resources from the deserts alone by 2020 – and potentially satisfy all of its energy needs in the longer term. At the same time, it could contribute to energy security in Asia, create a major new export industry and make feasible the domestic value adding of Australian mineral resources.

9

Desert democrats

'Negotiations between government officials and Aboriginal elders on culturally legitimate government arrangement had been proceeding steadily for months when, on the last day, a press release suddenly announced that the Australian Government would be taking over the administration of 60 communities. 'To say that the Bininj members of the … Committee were shell-shocked would be an understatement. In one day, without any consultation, their collaboration with the Australian Government had essentially been made null and void. Their role as the proposed local government for the entire region was thrown into question, their work over the last three years ignored, and their governance roles treated with disdain.'

SMITH 2007[194]

The brutal reality of living a long way from the centres of political power is that you are subject to the whims of distant politicians and bureaucrats and have little say in your own future. Your voice is a distant one, muffled by closer political thunder. From this distant voice arises social uncertainty and loss of local control in governance.

As we have seen, two of the desert drivers, remoteness and a sparse and mobile population, lead to the breakdown of mainstream models of governance and service delivery in the desert. Some writers, such as Mike Dillon and Neil Westbury, talk about there being a 'failed state' in outback Australia, and of Australian governments abandoning their obligations to govern the region properly.[195] We have shown how easily misunderstandings can also undermine the strategies that communities and

businesses need to adopt to deal with uncertain, scarce resources.

Despite these realities, excellent Australian models for better governance in remote regions exist. For decades, outback people have been demonstrating that they can organise themselves and their communities in ways that meet their needs and that mitigate the effects of indifferent and remote rulers. More importantly, the lessons learned in the outback can apply in other communities, including our mighty cities, as the need for all humans to live with resource scarcity imposes itself.

9.1 From aspirations to achievements

To turn the aspirations of voters into services that satisfy their needs, the job of democratic governments everywhere is to wrestle with four functions. Governments must get some sort of consensus from their constituents on their aspirations and needs for services, and devise a policy for delivering on these. They do this through processes such as community consultation and marketing a party platform at an election. Then they deliver the services, either themselves or by getting someone else to do it. Thirdly, they must regulate or police the quality of supply of that service, directly or indirectly. Last, of course, they must raise the money to pay for the service – or at least keep its costs within the funds they already have. At a national level, these service activities include defence and international trade, but here we are more interested in education, law and order, health, land, roads, rubbish, telecommunications, social welfare, and so on.

In reality, these four functions of governance apply at every level, from the individual household to the community and the nation. The matters at hand may be homework and house cleaning for the family, public toilets and rubbish for local government, land management and legal accreditation for state governments, and defence and international trade for the federal government. The processes of carrying out the elements differ at each level because of the different scale and the numbers of people involved, but the basic actions remain.

It is precisely on these problems of scale and population, of course, that the desert comes unstuck.

9.2 What's the problem?

The four elements of governance have to work differently with the desert drivers compared with what is possible in more populated regions. The issues that arise include autonomy (allowing people to express their local opinions), accountability (ensuring service delivery is held accountable to recipients), scale and cost efficiency (how to deliver services as affordably as possible) and clarity of roles (people tend to end up with multiple roles in small communities).

Desert people lack autonomy more acutely than small communities in populated areas. In particular, they are clearly remote from decision-making centres where policies are set that are intended to meet desert needs. The story at the start of this chapter illustrates how desert people can go about organising their governance arrangements – apparently with the blessing of governments – then find everything overturned by a sudden, distant and unexpected policy action.

This problem of distant voice affects all walks of desert life: Aboriginal communities, pastoralists, outback settlements, tourism operators and even mining companies. Social scientists Silva Larson and Yiheyis Maru made a study of what works and what fails in the governance of desert water resources in western Queensland and around Alice Springs. They concluded that central governments are happy to pay lip service to consultation, and place considerable responsibilities on the local community to engage in this.[196] However, they found that the resources provided for consulting were usually inadequate and local people were actually given little or no control of the final outcome. The ultimate say was reserved to the relevant Minister, who often felt little obligation to explain decisions. These decisions then ended up being filtered and altered by distant tiers of bureaucracy, often without explanation to the local people. Larson and Maru found that local people in the Lake Eyre Basin and Alice Springs were left feeling disenchanted and disenfranchised by what had started out as a promising process.

Even if a policy is set with local involvement, the resulting delivery of services to desert people is often unaccountable and ineffective. Here a key problem is one of scale and remoteness.

Where a settlement has more than a few households, it needs some form of organisation to sort out services. This is usually a local government or shire council for on-ground services, although other services such as education and power may be provided by a central government, and still others such as hairdressing and car servicing may self-organise through individual small businesses. Different countries have different norms about what each level of government delivers, so there can be a great deal of flexibility in how to arrange things, but we tend to follow our own cultural habits. For example, in Australia, fire services are centrally funded and coordinated by state governments, whereas in the USA every city block precinct may have its own fire service. It therefore comes as a surprise to Australians to find that US fire services can be poorly coordinated when it comes to big events such as plane crashes and fires on precinct boundaries. On the other hand, it ensures the local community has intimate knowledge of and loyalty to the members of their local fire brigade.

This creates a tricky trade-off. As the costs of services increase, the usual response is to seek economies of scale and have one organisation service more people for greater financial *efficiency*. However, to maintain local *accountability*, it is better to service fewer people, so that they are more closely involved in defining the services they want and in ensuring they are delivered satisfactorily. With rising costs and larger demand for services, this trade-off is a major issue everywhere today – as organisations get too big they fall out of touch with the people they are servicing. But, at least in a city, everyone lives near the service and the geography is small. In desert Australia, increasing your local council area in order to double its number of ratepayers may mean taking in an extra half a million square kilometres!

Economies of scale are gained by getting bigger, so that the fixed costs (such as of employing a CEO, having your own grader, and so forth) become a smaller proportion of the total budget, and the cost per resident serviced is reduced. For local government, economies of scale generally mean servicing more people – and this only works if the shire can acquire more people without massively increasing its area or the distances serviced. In the desert, this only works for a few services and it is necessary to be quite selective in choosing which ones, compared with urban areas. Even in cities, economies of scale only work for physical services such as sewage plants and rubbish collection, and not the 'people-intensive' services such as meals-on-wheels and childcare centres that are an increasing feature of local government.

The average Sydney suburb (and there are 650 or so of them) has a population of 6000 and spans a few square kilometres. The Shire of Ngaanyatjarraku around Warburton in Western Australia contains 2500 people and covers about 160 000 square kilometres.[197] Adding an extra suburb to an existing council in Sydney might involve people travelling a couple of kilometres further to reach their new council chambers. Doubling the Ngaanyatjarraku Shire population would mean people having to travel more than 1000 kilometres to reach the shire headquarters. (In fact the people of this shire already have to travel 1000 kilometres to the nearest District Court in Kalgoorlie.) This is not going to make rubbish collection any more efficient, and it is hard for people dispersed over 2000 kilometres to feel closely involved in the day-to-day running of their local council.

So, simply getting bigger in area is not the answer for most services. Even in mainstream local government, increasing attention is being paid to how economies may be gained through mechanisms other than increasing scale.[198] Besides economies of scale there are economies of *scope*. These deliver efficiencies by having one organisation, instead of several, deliver closely related services. For example, the health service, schools, police and local government may all need to consult with the community. Rather than each of them employing part-time interpreters and facilitators, one organisation can carry out these services on behalf of all. This drives service delivery organisations to engage in activities that are hardly the stuff of a typical local council. For example, the Ngaanyatjarra Council (which is not a normal local government) runs a food purchasing company, health clinics and a fuel supply business.

However, mixing up roles and functions to get economies of scope in community organisations can lead to another problem in small populations. In towns and cities, different functions can be separated into different sub-committees, or even different organisations, but in small communities the same people are involved in everything. For example, in an insightful study[199] of the tensions that arise when Aboriginal community councils are asked both to deliver services and to support cultural activities, Chris Adepoyibi found that such councils must interact with two quite different operating environments: one strategic, 'western' and formally structured; the other negotiating (and even protecting) the 'cultural field'. This creates endless tensions, a bit like mixing up church and state in European governments. 'A community council can be perceived by local people as a liberating force giving them the opportunity to achieve more autonomy and control over what they consider as of cultural and ceremonial value to them, on the one hand, and as a constraining influence, being an instrument for the state to exert and maintain control, on the other', he says. So, local people see the councils' role in maintaining culture as compromised. Yet, from the outside, the councils' essential role in delivering municipal services seems to be undermined by 'weak and irrational' decision making. No-one is satisfied. The conclusion of the study is that these sorts of roles should be differentiated – a conclusion supported by overseas cases.[200]

In fact, the expectation that the same body will broker the community's aspirations, develop a policy response, deliver the services and police their quality all at the same geographic scale also creates problems. Although brokering needs to occur very locally to ensure relevance and accountability, often the service may be better delivered at some wider scale for the sake of efficiency. Yet the tendency anywhere is for the governing agency – whether government, community or a company (especially if remote) – to want to totally control the service delivery, either by doing it all itself, or by creating a new 'trustworthy' body. Because the local delivery tasks are small in the small population centres of the desert, the result is a plethora of unsustainable, semi-competing, semi-monopolies struggling to survive in remote areas. In a more settled area, these may each have a large enough role that they can all be viable and deliver greater efficiencies through competition – but not in the desert.

Finally, desert organisations are beset by the common desert problems of the constantly changing community (see Chapter 3): staff turnover, small markets and high establishment costs (e.g. bringing in expertise from outside the desert). Yet they must access one-off, short-term grants schemes from funding sources split across multiple jurisdictions with differing application forms, timing and reporting requirements. Harry Scott was the town clerk of the small settlement of Titjikala in central Australia, which is home to about 250 people. As he says ruefully, he and the equivalent of one assistant had to manage 30 to 50 separate grants most of the time – covering everything from arts centres to child care, road maintenance to meals-on-wheels – in their total annual budget of around $2 million. Small settlements such as Titjikala have had to manage (apply for, budget, run, acquit, report, and so on) as many grants as a medium-sized town council, but each grant is much smaller and there are far fewer people to help with the administration.

However, as we know, the effects of remoteness and of distant voice are inescapable features of desert life; rather than wishing them out of existence, then, we have to look at the situation with new eyes. What policies and practices might make the process of governing somewhat better? And what can desert people do to insulate themselves from the inevitable reality that distant policy will pay little attention to them most of the time? As George Cooley, Aboriginal leader in Umoona in South Australia, says, 'Capable governance is essential – if you don't have it, educated people leave'. Desert people have found solutions to these issues, although their solutions are also repeatedly disrupted by distant policy makers who have no idea of the distances, issues and obstacles involved.

9.3 Remote area solutions

As we have seen in Chapter 7, the survival of desert settlements – whether pastoral, mining or Aboriginal – depends on their services meeting the aspirations of their community and being affordable. This means the community needs to be closely involved in defining what services they want and what they can afford, and in developing creative ways to

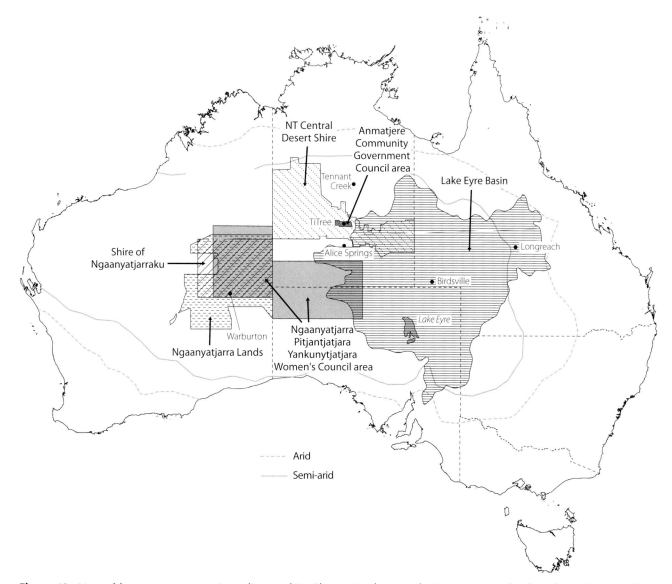

Figure 13: Map of key governance regions discussed in Chapter 9, showing the Ngaanyatjarra lands and roughly coincident Ngaanyatjarraku Shire, the Anmatjere Community Government Council and the Central Desert Shire that has superseded it, the Lake Eyre Basin and the Ngaanyatjarra–Pitjantjatjara–Yankunytjatjara (NPY) lands over which the NPY Women's Council operates (see Box 18).[201]

deliver those services. Here lies the challenge of governance in remote areas; and there are success stories from which we can learn.

Perhaps the most notable example of remote service delivery and self governance comes from bottom-up innovation on the Ngaanyatjarra lands in Western Australia (Figure 13) over the past 30 years. These lands consist of 13 separate Aboriginal settlements, each with a store, football oval, waste dump, power generator, water supply and other

local facilities. Each settlement is incorporated as an independent Community Council, and is the principal home of between 80 and 600 people (who are also highly mobile among settlements), so they are generally too small to support their own essential services. In 1981, the settlements banded together to establish the Ngaanyatjarra Council[202], which, since then, has provided major services such as health, air transport, fuel delivery, store supplies and accounting, among many others (see Table 4, Figure 14). The

Table 4: The conglomerate of service delivery businesses based in Perth, Warburton and Alice Springs that the Ngaanyatjarra Council was overseeing in 2003[a]; more recently a Land Management Unit was added, but the airline is no longer operational.

- Ngaanyatjarra Health Service
- Ngaanyatjarra Services Accounting and Financial Services
- Ngaanyatjarra Services Building Division
- Ngaanyatjarra Services Works Division
- Ngaanyatjarra Air – Airline/Charter Service
- Ngaanyatjarra Agency and Transport Services
- Native Title Unit
- Ngaanyatjarra Community College
- Ngaanyatjarra Media
- Ngaanyatjarra Council Savings Plan
- Ngaanyatjarra Marshall Lawrence Insurance Brokerage
- Ampol Alice Springs (Indervon Pty Ltd).

[a] <www.id.com.au> (accessed March 2009) and Ngaanyatjarra Council (2003)

Ngaanyatjarra Council consists of the chairs of each of the 13 settlement councils, and operates through a small head office run by a coordinator. All people in the lands can vote for the chair of the Council, so there is a sense of direct bottom-up democracy about its operations.

Although it has had its ups and downs, the Ngaanyatjarra Council persists today as a successful example of how local people can organise themselves into a structure that combines both regional and very local elements. The settlement councils manage local day-to-day activities, such as sporting facilities, waste disposal, their power plant and the

community store, which can (and need to) be performed efficiently in each settlement – incidentally providing local employment. Meanwhile, the wider regional body takes on those services that are inefficient at the small scale, such as health services, fuel delivery and stores procurement. It also gains economies of scope across the region, with one head office, one coordinator and one administrative team, as well as opportunities for synergies between services, such as the health service using the airline. Being independent of any formal structures, the Council is very flexible. It has a main office in Alice Springs (as well as Warburton), but, for example, the stores procurement and some legal work is done more efficiently from Perth. Fuel delivery is not very efficient even at the scale of the whole region, and this arm of the Council also delivers fuel into neighbouring regions and runs a service station in Alice Springs to be viable.

An interesting side effect has been that the Council decided early on that it should not deliver services to any settlement too small to support at least a small store. Consequently, the issue of very small outstations unsustainably subsidised by public funds (although, of course, created by distant government policy in the first place) that has plagued other regions has simply not arisen in the Ngaanyatjarra lands. Local decisions have had sensible local outcomes – no settlements that are judged 'non-viable' by the locals (see Chapter 7).

This is a living expression of the European Union's principle of 'subsidiarity', where only functions that cannot be performed efficiently and effectively locally are passed to a more central body. The fact that the local bodies are really in charge (Figure 14) provides a strong democratic sense of accountability between the levels. However, the Ngaanyatjarra Council is not established under any local government legislation – indeed, the scope of its operations would be a bit hard to imagine a typical local government doing (e.g. buying food supplies for local stores!). In fact, the Shire of Ngaanyatjarraku, which was established under Western Australia's Local Government Act, operates across roughly the same area (i.e. about 2500 people in 160 000 square kilometres). Ngaanyatjarra Council

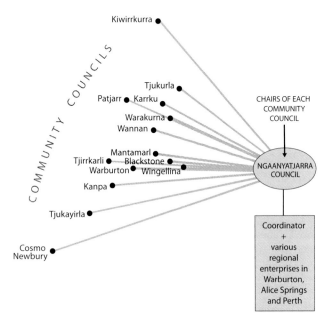

Figure 14: The 13 Ngaanyatjarra communities that underpin their subsidiary body, the Ngaanyatjarra Council.[203]

BOX 20: LANDCARE AND NATURAL RESOURCE MANAGEMENT GROUPS

During the 1980s and 1990s, Australia created a new tier of community-level natural resource management through the Landcare movement. People banded together locally to deal with issues that occurred at a broader scale than the individual farm, such as catchment management of weeds and feral animals, group learning about rehabilitation and water harvesting methods, and so on.

During the 1990s, central government realised that the Landcare movement had been very successful at mobilising local action, but this was occurring without a coordinated approach to regional priorities. Although some projects were excellent, in the end they were rather *ad hoc*. The new National Heritage Trust therefore pressed for regional-scale planning and implementation, and 56 regional bodies aimed at 'integrated natural resource management' were set up across Australia. It was left to the states and territories to work out how these should be established in each jurisdiction – as long as they met a few standard criteria. The result was an interesting experiment – Queensland established community groups that saw themselves as fairly independent of government; South Australia and New South Wales created statutory boards – in New South Wales, even taking over many roles that were previously departmental. The Northern Territory established a single body for the whole sparse population, and Western Australia set up a single body for the vast rangelands area of that state.

The different regional bodies have made great headway in establishing accredited plans for their regions, and channelling public funds through these plans to projects, some of which are then run by the original Landcare groups. However, two issues are apparent. Firstly, the current regional groups are not community created: each state in its own way took centralised decisions about their bodies. This created experimentation between the states, which has been useful, but has limited the level of local ownership, which has not. Secondly, Ministers or government 'joint steering groups' retain much of the final say over resource allocations; several analyses[204] show that management *responsibilities* are devolved to the groups to a much greater extent than the matching *rights* and *resources* required to fulfil these responsibilities. Thus the approach has made decision making about priorities more regionally relevant, but has also disempowered the very local level. A better balance between these is required, which would involve the local groups formally constructing their own regional body.

tenders competitively for a number of services needed by the shire, including road maintenance and some building and accounting. Naturally, in such a small community, there are strong personal links between the organisations, but their functions clearly differ.

There are many other governance lessons to be gained from organisations across remote Australia. The development of Landcare, then the more recent regional natural resource management groups, has been an effort to devolve governance of resource management services (Box 20), with some degree of flexibility in arrangements in different states; however, the devolution of responsibilities has not really been matched by that of rights and resources. In the Lake Eyre Basin, across Queensland, South

Australia and the Northern Territory, community efforts to coordinate water management created a vibrant bottom-up Coordinating Group; however, this was overtaken by top-down national developments, with some positive outcomes but also some limitations (Box 21).

It is not easy to make general rules about the appropriate scale for governance arrangements, as illustrated by the changing nature of institutions in the Anmatyere region (Box 22). However, the latest version of these is the enormous Central Desert Shire, which is constructed from the top down rather the bottom up so that it is likely to struggle to stay sensitive to local conditions. Outside the desert, the Torres Strait Regional Authority is an arrangement reminiscent of the Ngaanyatjarra Council,

BOX 21: NATURAL RESOURCE MANAGEMENT IN THE LAKE EYRE BASIN[205]

In 1995, the pastoral industry, conservation groups, mining and petroleum industries, Landcare groups, Aboriginal organisations and local government – as well as Queensland and South Australian Governments – formed a group to protect the future of the Lake Eyre Basin (LEB). Initially triggered by concerns about World Heritage listing, this move was sustained by opposition to damaging cotton farming developments on the Cooper River at Currareva. After consultations across the immense area of the Basin (Figure 13), from Longreach to Alice Springs and down to Marree, a series of 'local' catchment management groups were set up, which were not constrained by state borders. Members of the catchment management committees formed an over-arching Lake Eyre Basin Coordinating Group (LEBCG), with some additional members from government to maintain connections and some independent experts as a source of knowledge. The LEBCG and some catchment committees were established in 1998.

Meanwhile, the process had precipitated moves towards an LEB Intergovernmental Agreement between Queensland, South Australia and the Commonwealth. This became operational in 2001 and was joined by the Northern Territory in 2004. The Agreement established a Community Advisory Committee, which took over some of the LEBCG's role, so the latter was mothballed in 2003. At the same time, regional natural resource management groups were established under the federal Natural Heritage Trust from 2001 (see Box 20). These could not operate across borders, so the Agreement became all the more important to maintain such coordination.

The LEBCG was essentially a community-driven, bottom-up process. It was focused on natural resources, especially water, but the focus was interpreted widely and its membership included local government and industry. The LEBCG was generally subsidiary to the catchment groups. That this small community group triggered an intergovernmental agreement was a massive achievement in one way, which recognised the fundamental fact that water does not respect human boundaries. However, the formalised Agreement narrowed the scope of concern to water resources only – no longer operating from the bottom up – and reduced the flexibility of community input (through its advisory committee) to consider wider regional development issues. At the same time, the distant policy changes that established regional natural resource management groups had a big impact on the local catchment groups; fortunately, the key regional bodies (Desert Channels Queensland and South Australian Arid Lands Board) have continued to support the catchment groups and try to work across the borders[206], but they might not have. The LEBCG was only mothballed, and one day it may rise again!

being a regional council made up of the chairs from its 18 Island Councils, plus two separately elected members, which is widely acknowledged for its effectiveness.[207] However, the Authority is established under federal legislation, while the Island Councils operate under Queensland legislation, and have been caught up in local government changes across that state. This again highlights the fragility of remote area arrangements legislated by distant governments.

These practical case studies demonstrate that is possible to govern remote areas well by centralising some services across the whole region and devolving others to local settlements. The balance should be chosen to make local sense in response to the desert drivers of remoteness and sparse populations – the scales of governance and of service delivery do not have to be the same. In fact, to gain efficiencies, organisations operating at any given scale should go mostly for economies of *scope*, not of scale. However, the earlier examples show that it is really important not to mix roles such as cultural and administrative obligations.

Furthermore, the Ngaanyatjarra Council and the original Lake Eyre Basin Coordinating Group were constructed from the bottom up – local organisations

BOX 22: ANMATJERE COMMUNITY GOVERNMENT COUNCIL[208]

It is hard to generalise about the right scale for regional arrangements because people's attitudes and service delivery costs change with time and context. Social researchers Will Sanders and Sarah Holcombe studied the Anmatjere Community Government Council (ACGC) north of Alice Springs. The ACGC set itself up in 1993 as a multi-settlement regional local government with nine wards based around outlying settlements. Many of these settlements maintained their own incorporated associations, running key local services such as Community Development Employment Projects. It was another form of the arrangements that were running in the Ngaanyatjarra lands and the Torres Strait. The ACGC serviced a total population of about a thousand in settlements spread across 15 000 square kilometres of country.

During the early 2000s, the ACGC gradually centralised activity towards its headquarters in the village of Ti Tree, through a mixture of staffing, funding and policy pressures; only the outlying wards that had active settlement councils seemed to stay independent. From Sanders' description, it nonetheless seems that the original scale was about right for the time.

However, the ACGC was subject to local government legislation, which changed in 2008. The Northern Territory moved to greatly increase the scale of regional governments. It created a Central Desert Shire[209] with four wards to represent and service about 3800 people – about four times as many settlements as previously, and over 40 times the area. The new Shire is at least 1500 kilometres across from its north-west to south-east corners, larger even than the Ngaanyatjarra lands. The legislation permits local boards to be set up, but they have no formal role in governance and local people may feel little sense of engagement with the new entity. It will require a major effort to sustain localism within the imposed distant regionalism of the new arrangement. Once again (see Box 20), a better design would be that the local groups had been able to set up the regional body, rather than the other way around.

were there first, and they agreed to create a more regional body to do the things they could not do locally. Other examples were constructed mainly from the top down – some sort of regional body was established, which later chose whether to have local groups formally included. Given the large distances between local groups in the desert, this seems to reduce their ability to represent local differences.

However, the legal status of the different case studies presents a conundrum. Ngaanyatjarra Council is outside the formal state-legislated local government arrangements. The Torres Strait and Anmatyere arrangements were within these, and both have been affected by distant, one-size-fits-all policy changes. The Lake Eyre Basin moved to greater security with its Intergovernmental Agreement, but with the benefits of that came a narrowing of agenda. The conundrum is that being formally legislated should make

arrangements more reliable, but, in practice, unconventional, locally controlled arrangements seem to survive better in remote regions. This statement has to be tempered: most of the case studies have faced the usual desert issues of lack of capacity, staff turnover, short-term contracts, and the fragile role of leadership. Hence it is necessary to be realistic about the capacity of very local organisations to carry out all the necessary roles.

So can we do better? In fact, there is a good basis of research on all these issues to build on today.

9.4 Designing better forms of governance

Local arrangements usually only get undermined when the community moves into closer contact with the pressures of distant markets and governments that do not know, and may not care, about local conditions. For two decades, American political scientist Elinor Ostrom has inspired a group of researchers around the world who have studied how many local communities manage their own communal resources[210] – whether they are pastoralists or fishermen or farmers. It turns out that they create their own management arrangements, such as local fish catch limits or water rights, which are very effective at conserving the resources in the long term. There are now well-understood and systematic 'rules about the rules' for such institutions. There are seven categories of rules, including who is covered by the arrangements, what their rights are, how group decisions will be made, what sanctions may be taken against individuals if they transgress the rules, and so on.[211] In traditional societies, the rules may often be unspoken norms, but it is better that they are explicit if there is population turnover with outsiders coming in, so there is no misunderstanding among naïve newcomers. Some of the rules relate well to what the people in our success stories have worked out for themselves. Sparse populations and social uncertainty mean that it is particularly important to be clear about many of the rules in deserts.

Ostrom also shows why a series of relatively independent, self-organised resource governance bodies may do a better job of regulating common resources than a single central authority. Among the benefits are: more effective use of local knowledge; a better understanding of who is trustworthy (greatly reducing the costs of enforcement); locally appropriate adaptation of rules; and redundancy, such that, if the system fails, it only does so locally. There are also potential faults with local organisations, of course: in small, remote populations, they can be corrupted or stagnate, they may lack the capacity to develop new information, they may not learn, and they may be wrongly scaled for the purpose in hand. Hence, simply devolving *all* authority to very local organisations is not generally the best solution. Ostrom argues instead for a system where citizens are able to organise at more than one scale.[212] Regional bodies can then provide a context for local organisations, which may differ in detail between locales. This is exactly what the Ngaanyatjarra Council has done

(see Figure 14). The regional bodies can also play the important role of assisting learning among local organisations as they experiment with the best arrangements. Ostrom remarks, 'Such systems look terribly messy and are hard to understand. The scholars' love of tidiness must be resisted!' So, indeed, must the love of administrators for universal and centrally controlled solutions, which is what has happened with the natural resource management bodies (Box 20) and most local government legislation (e.g. Box 22). It *is* possible to set the key rules simply and universally, and then allow locals to experiment in how to implement them.

Mark Moran and Ruth Elvin's research on community governance in remote Australia provides another angle on these design issues.[213] They argue for three key characteristics that relate to Ostrom's design rules: subsidiarity, connectivity and accountability. *Subsidiarity* holds that government should undertake only those initiatives that exceed the capacity of individuals or private groups acting independently. In other words, the higher level of organisation is subsidiary to the lower ones, not the other way around, as is usually presumed in hierarchical systems. It is often stated (most famously in the European Union) as being the principle of making decisions as close to the level of the individual citizen as possible. *Connectivity* refers to all the connections and networking in a governance system between individuals and groups. There is 'horizontal' connectivity between groups and communities in the desert; and 'vertical' connectivity between the desert and levels outside, which tends to be much weaker. *Accountability* refers both to the duty to report upwards to a demanding big government elsewhere, and to the duty to report downwards to local stakeholders, who may often be widely scattered. Upward accountability is in the ascendancy at present, which adds greatly to the administrative load on small organisations.

These characteristics are particularly important in remote contexts. Subsidiarity is important because there is a huge physical disconnect between 'local' and the next scale up in remote areas. Connectivity is particularly important because organisations are usually remote from one another and it is difficult to network well. And it is crucial to get expectations around accountability right – where there are few people, the paperwork involved in reporting upwards can be overwhelming; and being accountable downwards is hard if the people you need to report to are scattered across a wide area. Getting the balance right with these design characteristics leads us to the practical side of governance.

Understanding how to make engagement between government and the community work well has been the focus of work in the Lake Eyre Basin by social geographer Tom Measham and his team. They talked to residents to compile a list of successful operating principles for delivering natural resource management in remote areas. In summary,[214] they say that successful engagement relies on maintaining community trust, while carefully crafting and navigating governance processes. This means understanding local culture and needs, as well as those of government policy objectives, and finding creative ways to resolve the differences.

Echoing the experiences of our success stories in the previous section, they found that good outcomes depend on good networks that build trust and transparency in these sparsely populated areas. People play multiple roles in small communities, so it is vital to be clear what 'hat' is being worn when. Face-to-face communication is best, but can be expensive; however, modern technology can help maintain the links, as we saw in Chapter 6. It is also vital to support local desert champions, because individuals are particularly critical to success in the sparse populations of remote areas. It is vital to identify, train and retain these individuals in the community and in local organisations. Lastly, Measham found that 'local desert time frames matter': thinking ahead, being prepared to take advantage of unpredictable (and infrequent) opportunities and maintaining commitment are crucial to long-term survival. It also takes time to communicate, network and build trust.

Measham's findings resonate well with those focused specifically on Aboriginal governance. Australian National University researchers Diane Smith and Mick Dodson have led a team that has been exploring features of successful Aboriginal systems of social and political organisation, which draw from their long history and cultural milieu but are evolving to deal with today's conditions. These principles are not particularly 'Indigenous', nor all remote – they are good principles for governance in general. 'They could usefully inform the governance strategies and action of other communities and organisations', says Smith.[215] Approaches that are particularly relevant to remote areas readily emerge from their work. Some of their principles echo issues of subsidiarity and scale – particularly balancing local and larger-scale representation, and emphasising relatively egalitarian relationships among organisations. They find a strong emphasis on connectivity and networking, both within and outside organisations and their regions. And they highlight the need to build the skills that are often lacking in remote regions: among key leaders, their staff and their institutions.

In short, there are plenty of theoretical understandings and practical lessons around from a variety of arenas with clear common themes. Surely we can now set out the most important principles to make desert governance work much better than it does today.

9.5 Desert design principles for the future

Australian governments are failing to apply these approaches to desert governance. If anything, they are adding to the sense of isolation felt by desert people – whether Aboriginal or not – when remote power brokers either ignore or dictate to them. Such actions undermine the rights of remote inhabitants, while adding to their responsibilities. They also keep the inhabitants in the dark about the justification for decisions that affect their futures, denying them the scope to challenge those decisions. This is fundamentally undemocratic.

Local government reforms that involve merging shires are also being driven by models from closely settled regions. These models are the precise opposite of subsidiarity, which aims to devolve decision making as close to the community as possible and creates larger bodies only where there is a need. Most of Australia's regional bodies for natural resource management have also been established top down – yet it is the people on the ground, in the end, whose actions and decisions will decide what happens to the landscape. Small wonder then that people are talking of inland 'failed states' and of government abandoning its responsibility to govern outback Australia![216]

The desert's distant voice will always condemn it to cycles during which it is first ignored by governments (whose natural focus is mostly on the main population centres) and then becomes an unwelcome centre of attention for interventions that may not work as intended under the unique conditions that prevail in the deserts. The point is not that governments should pour money into desert Australia, or even that they are remiss in forgetting about it from time to time. The point is that better forms of governance are required, which meet local needs and reflect national policies, and which can ride out periods of neglect by centralised authority: governance that is resilient in the face of how the system naturally works.

From the practical examples and theory reviewed above, the following six essential principles for governing remote regions emerge:

1   Devolve as much responsibility as possible to local levels, so that government is most answerable to the people it serves, and allow these levels to define any broader regional government bodies (subsidiarity). In remote urban centres such as Alice Springs, this means the current style of arrangements could continue. Outside these centres, settlements with a distinct community of interest – based on locale or livelihood (see Chapter 7) – would elect their own settlement council. These councils would then send a member (probably the chair) to sit on the shire, and the shire would be seen strictly as a subsidiary body to the settlement council level.

2   Greatly diversify the role of regional shires (e.g. involve them in natural resource management, language centres and other currently unconventional roles) to reduce the costs to small populations of having too many separate administering organisations. The regional shire should encompass the needs and aspirations of all the communities of livelihoods (see Chapter 7) contained within it. Remote areas simply cannot afford the multiple governance systems that are typical of densely settled areas. The settlement-specific bodies need to be able to retain autonomy of structure and function to suit their settlement type. Their main obligation would be to be inclusive. There remains a question as to whether central government should legislate for this scale, or simply facilitate its creation informally by the local bodies; we favour the latter approach because it permits regions to work out their own needs. On the Eyre Peninsula, for example, the regional development corporation, regional local government association and the regional natural resource management group have already started to coordinate activities spontaneously and flexibly.[217]

3   De-couple service *governance* by all the tiers of government from service *delivery*. This means clearly separating the role of local government in determining local aspirations, setting service delivery policy and monitoring outcomes from that of actually delivering the services (which can often be done by a community body or small business). This is a critical principle for enabling the governance bodies (principle 1) to operate more locally than efficiency would otherwise allow – the policy setting and accountability functions are far cheaper per head than the service delivery ones. At times, local government might deliver services too, but only where it makes good economic and social sense to do so. At the same time, local settlement councils should not be barred from banding together to establish a commercial entity along the lines of Ngaanyatjarra Council. These commercial bodies should be able to tender to any legislative structure and be large enough to outlive policy changes, and should not generally have a one-to-one relationship with local government bodies. The choice of scale will depend on purpose.

4   Avoid inappropriate mixes of roles in one organisation – in particular, keep cultural responsi-

bilities distinct from civic ones – while accepting that the same individuals may sit in different roles in the different organisations.

5 Invest in good networking, information flows, connectivity, valued individuals and leaders, and the use of new technologies in all these bodies.

6 Decrease the complexity and wasted effort surrounding how local bodies apply and account for funds from higher levels of government.

The last three principles can be further enriched by the work outlined in the previous section. One should not be naïve – in reality, these will always be under challenge from the desert drivers, and particularly by the nature of remote decision making. However, with an improved structure of local and regional governance, remote regions will become more resilient to these effects.

The principles are framed in terms of local government in particular, but actually apply in various ways to many organisations trying to operate in remote regions – distributed government departments, large corporations such as mining companies, non-government organisations, large pastoral companies, and so on. All need to operate better in remote regions.

9.6 Resourcing desert Australia

In balancing responsibilities, rights and resources, it is particularly critical that regions are properly resourced, and have genuine rights over those resources. As noted in Chapter 1, in recent years, desert Australia has generated over $90 billion of gross revenue for Australia every year – nearly a twelfth of the nation's annual gross domestic product – and it has exported literally trillions of dollars of value since Europeans arrived on the continent. The desert also exports human capital – people – and natural capital in the form of water and precious nutrients embedded in beef and other products. A very tiny proportion of this has been re-invested in desert capital (whether human, social, infrastructure or the environment). In fact, most

publicly funded programs live a hand-to-mouth existence, with never more than a few years' assured funding. This is the reality of the desert drivers, in that the active interest of the outside world is never assured. So, what actions at the scale of desert Australia as a whole would best improve the resilience of the region in the long term?

The answer is a structural change that would be hard to enact but which, if established in an opportunistic moment of sympathetic support by the rest of the country, could provide some consistent and secure resourcing for desert Australia. It is an Outback Capital Trust Fund, established through collaborative state, territory and federal legislation. This would have the power to levy rents on all uses of natural resources in the outback. Its trustees would obey a charter to invest the financial capital from those levies to the best effect to improve the state of the natural, social, human and physical capital of the region. The trust beneficiaries would be defined as all outback inhabitants. A variety of conservation management, Aboriginal, social, communications and transport projects would be funded by the Trust, aimed at increasing the levels of these different capitals.

Is this a pipedream? No, it could be set up rather like the Alaska Permanent Fund, which has been operating there for 30 years, and there are other examples around the world.

The Alaska Permanent Fund[218] was established by a change in the Alaskan Constitution in 1976 as 'a means of conserving a portion of the State's revenues from mineral resources to benefit all generations of Alaskans'. It was to receive at least 25 per cent of all such revenues, but has mostly received 50 per cent. It now exceeds private funds such as the Paul Getty Foundation in size, being worth $36 billion in 2008 after earning about $2.9 billion that year (compared with only 0.84 billion in actual mineral revenues, which is what the state would have received in the absence of the Fund). The earnings include a dividend paid to all Alaskan residents (total $1.3 billion in 2008), with most of the rest going into general revenue for use according to the elected legislature's priorities.

The Alaskan Permanent Fund is overseen by a board of six trustees appointed by the Governor, mostly with 4 year staggered terms. The Board is politically neutral, and the fund is 'insulated (but not isolated) from political activity'. Capital cannot be withdrawn from the Fund without a constitutional amendment. The dividend component disproportionately raises the income of low-paid Alaskans, especially in rural areas. The capital base has the effect of greatly reducing the boom-and-bust nature of oil and gas income in the region, thus reducing the variability in the benefits. The general revenue component remains available for use within the local (i.e. Alaskan state in this case) democratic process.

There is a strong case for establishing an Outback Capital Trust Fund, with a similar goal of retaining a component of capital funds earned from resource use in the outback to support human, social, physical and natural capital investments, which are otherwise lost to the region. As is evident from the Alaskan case, it is critical to construct such a fund carefully; this is not a model like the existing Aboriginal Trust Benefits Fund, nor a Future Fund, which can be raided relatively easily by the Treasurer. It should be legislated jointly among the desert states, Northern Territory and Commonwealth to have an independent board appointed by government to oversee the fund, but with long terms for each member. It should pay dividends to all desert dwellers, but have most of its funds disbursed through regional governance processes with proper accountability to desert people. It needs to be established in such a way that no government can raid the capital base, and it might have a requirement to spread its benefits among the different forms of capital.

The Outback Capital Trust Fund would focus on rectifying the under-investment in desert *capital*. Parallel to this, administrators Mike Dillon and Neil Westbury have proposed some modest, but far-reaching, changes to transparency of the Commonwealth Grants Commission processes, which would help to fix up the extent to which intended *recurrent* government spending never reaches the desert (Box 23).

If the Outback Capital Trust Fund was established, and the Grants Commission process was altered as suggested by Dillon and Westbury, the future for the three-quarters of Australia that is desert, and that of all its inhabitants, Aboriginal and non-Aboriginal, would look much more secure.

9.7 Governing beyond the deserts

As rural, remote and regional areas become more alienated from the world's huge urban power centres, calls for devolution of governance are growing stronger worldwide. Australia is grappling with this challenge – but so is almost every country, especially those with large hinterlands.

Governance may seem like a dry issue, but it goes to the heart of one of the most important challenges facing humanity – our ability to feed and sustain ourselves on a finite planet, through the peak of human population and resource demand. The people who inhabit rural regions are stewards of most of the world's productive land, fresh water and biodiversity that are the essential life support system for the majority who live in cities. Rural people need the recognition and support to manage these resources well on behalf of the whole of humanity. Only then can they respond sensitively and flexibly with the local strategies needed to handle future variability – building reserves, protecting refuges and perhaps maintaining mobility in the face of uncertainty.

Despite often good intentions, most current models of governance for rural areas do not work well. Far beyond the deserts, the failures of central government to genuinely relinquish control and allow bottom-up systems to work are lamented.[219]

Issues that are rising on the agenda all over the world are brought into stark relief by the extremes of the desert, because the effects are so much more exaggerated there. The following lessons will help rural areas outside the desert, right across the planet:

- Devolve power to local levels by implementing subsidiarity; create local government that builds from the settlement scale upwards, not

BOX 23: OPENING UP THE GRANTS COMMISSION[220]

Mike Dillon and Neil Westbury in their book *No More Humbug* have proposed an elegant solution to the leakage of recurrent funding from remote regions.

At present, the Commonwealth Grants Commission makes calculations each year of how much funding is needed in each state (or territory) to deliver an equitable level of services. In doing this, the Commission takes account of the extra costs of operating in remote areas, using a sophisticated set of calculations. On the basis of these, funding is passed over to each state. However, the state is then free to spend the funds as it sees fit, which is reasonable given that each has a democratically elected government. The problem is that the desert drivers ensure that less funding ends up being spent in remote areas than the Commission's calculations would suggest is needed. For example, one study[221] showed that only 26 cents in each dollar allocated to remote area education actually make their way outside Darwin in the Northern Territory – the rest ends up being used for administration in the city, or fails to be spent at all because of recruitment difficulties.

Dillon and Westbury's simple solution is to suggest that the Commonwealth Grants Commission should explicitly indicate how much of a state's income was targeted at the remote regions, and the states should be required to report how much they actually spend there. The state still has the right to spend the funding as it sees fit, but there is now transparency and its populace can decide whether they think the spending pattern is fair or not.

This would be an excellent solution to improving the recurrent expenditure for remote regions, but does not tackle the issue of capital investment (which, in the long term, determines how much recurrent funding is actually required). For this, another approach, such as the Outback Capital Trust Fund, is required.

from the region downwards, and with only limited central guidelines aimed at accountability and lawfulness.

- Capture economies of scope by delivering health, education, natural resource management and other services in combinations and at the scale that makes the most sense in a particular region – not in a 'one size fits all' approach.

- Enable local and higher levels of governance to connect more thoroughly and powerfully with one another, especially if they are dispersed over large areas.

These principles are not only about governing rural or remote regions but about structuring the global society for the future. Multinational companies also run increasingly remote and dispersed enterprises, as do non-government organisations.

More broadly, society is becoming both more connected and more fractured, with the internet allowing the development of communities of ever-narrowing interests who may not even live in the same country, let alone the same street. At the other extreme, some people are being rendered ever more powerless by being cut off from technology and knowledge – creating 'hinterlands of neglect', even within our cities as well as beyond them.

In the face of these changes, our democratic structures cannot cope with the rising power of special interest groups. Our politicians are driven more by opinion polls than by the common good. Our systems of representative democracy are distorted by powerful, but corrupt, individuals and organisations.[222] Society is increasingly demanding new models of participatory and deliberative democracy, where strategic discussions can be had more thoughtfully, and aspirations can be worked through more realistically. Problems that need solving at the global level, such as climate change, add complexity to the challenge.

It is no overstatement to say that democracy in an age of globalisation and global change means that ordinary citizens have to become more involved in governance. Yet our conventional forms of government are failing to find adequate ways to devolve these rights and responsibilities. The problem is particularly apparent in remote regions, but so too are the solutions.

10

Deserts and our future

'This land must change ...'

MIDNIGHT OIL[223]

Tomorrow's desert sun glints over the rim of an ancient red range. A timeless wave of sand dunes stirs in the hot wind that forecasts another scorching day. Visitors wilt, but desert oaks whisper on the breeze, military dragons lift their legs in casual salute and desert dwellers settle on their haunches in the ways of a thousand years. They are not just survivors: these inhabitants thrive on desert living, making the most of extremes that, unheeded, destroy. This is no foreign fearful land to them; yesterday and tomorrow, this is a hard, but happy, home.

But today is less certain.

Despite innovation and goodwill, desert people are divided and distant administrations are frustrated by their inability to make desert Australia flourish. Despite the best of intermittent policy intentions, things are not right for Aboriginal people, for pastoralists or for other inhabitants of the great inland: people are leaving many desert regions, its pulsing heart is febrile and all the riches of the mining boom seem powerless to reverse the trends. Yet the desert remains a vital supplier of resources to Australia, as well as a repository of natural and cultural heritage and, indeed, a source of national identity. It remains a great place to live, and a place where the nation needs people to live: to manage the land, provide security for visitors and support the resource industries. From a better understanding of how the region works emerges a real blueprint for the future of this red land.

And this is no longer simply an issue for the desert interior – that fabled remote outback that comprises three-quarters of this southern land. Not so long ago, deserts extended to the coast on three sides of the continent; desertification and climate change mean they are on the march again. The essential nature of Australia is reasserting itself. Answers to the challenges of scarcity, uncertainty and aridity are increasingly needed for the rest of the country, as well as around the world.

10.1 Living with the desert syndrome

In Chapter 3, we formally introduced the idea of the desert drivers, arguing that they are causally linked in desert conditions. These were an unpredictable climate, scarce and patchy resources, sparse populations and mobility, remote markets and isolation from political power ('distant voice'), the importance of local knowledge, cultural differences and social uncertainty.

Throughout the book we have seen many examples of these drivers in action. Plants, animals and the pre-colonial Aboriginal economy were driven by the variable, unpredictable, scarce and patchy nature of their water and nutrient supplies. They all developed strategies to survive: being locally persistent, refuge dwellers, ephemeral, nomadic, or exploiting the desert from outside (Plates 1–3). We also saw how organisms can team up so that one facilitates the establishment of another, or create self-organising communities that can manipulate the flows of these limited resources (Plate 4). Aboriginal traditional society in particular adapted to use most of these strategies at one time or another, developing intense systems of local knowledge to support them in so doing. This knowledge was not only 'about the environment' – as western thinking tends to put it – but it also taught people how to interact with that environment, and how to self-organise to do so most effectively. These are elements, it must be said, that are still missing from the way that many urbanised European and Asian settlers interact with Australia, its landscapes and waters. Yet they are the keys to our common future, and the failure to understand them condemns us to undermine them through their weak points (Plates 5–6).

Although the biophysical drivers mattered most in the past, today it is the other desert drivers – a sparse and mobile population, remoteness from markets and decision making, and the resulting social uncertainty and differentiated culture – that play a bigger role in modern desert living. Many aspects of desert life still have to deal with sparse and unpredictable resources in some form, so the biological and Aboriginal ways of handling this keep reappearing as alternative response strategies. Desert livelihoods and small businesses, desert settlements and their services, desert governance and institutions – all must deal with a greater diversity, and perhaps magnitude, of uncertainties than most of their counterparts in the more populated regions. However, they must also deal with issues of critical mass, mobility, physical remoteness and distant voice. As we have seen, this has spawned both innovation and difficulty (sometimes disaster) for life in desert Australia (Plates 6–8).

The desert drivers are not to be taken lightly: a conclusion we drew in Chapter 3. Now that we have explored their implications, it is possible to see that they tend to limit the conventional livelihood options open to people, and to cause money and people to be drawn out of the desert. Left unmanaged, these effects result in a feedback loop that keeps the population low and sparse; we call this cycle the 'desert syndrome' (Figure 15).

The desert syndrome is self-reinforcing and it is pointless to rail against the desert drivers themselves. However, just as pastoralists and pre-settlement Aboriginal people came to terms with the inevitability of droughts and floods, and learned to manage *for* them instead of *against* them, so other desert dwellers need to come to terms with the effects of the desert syndrome. Governments will inevitably focus elsewhere – because that is where most of the people are. Markets will inevitably be small – it is just a characteristic of the desert. Social networking will inevitably be harder – it is just the result of being a remote, sparse population.

These conclusions are no cause for despair – they are a call to action! Each desert driver creates as many opportunities as it closes off (Plates 7–8). Distant government can also allow greater freedom for local interpretation. Small markets offer the opportunity to test products that would not be viable elsewhere. The difficulty of networking creates revolutions in transport and communication. The desert syndrome creates a context for all these things, which is different from that of urban areas. It is not a better or worse context than that of urban areas, just *different*!

Problems only arise when we try to enforce urban norms on the desert context. Sometimes this happens because people coming to live in the desert expect the same urban norms to apply. This can be fixed by the desert dwellers themselves understanding how the desert really works and explaining this to newcomers. At other times, outside decision makers set up the business, service and administrative context within which desert dwellers have to operate in ways that clash with desert needs. This is common and more awkward to fix, precisely because of the desert syndrome and the difficulty of influencing distant decision making. This book is seeking to raise awareness about these issues.

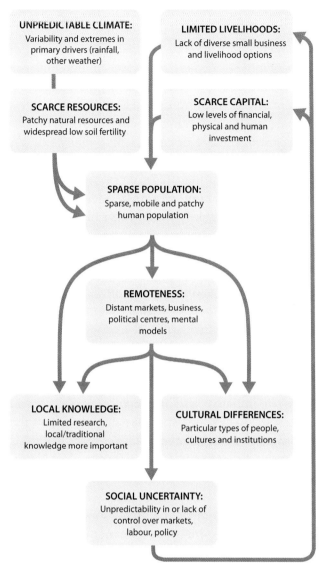

Figure 15: The desert syndrome created by the desert drivers in Figure 4, identifying key aspects (top right) of a feedback loop that tends to keep the desert population low and disadvantaged in a world that is inevitably driven by the interests of its population centres.

Awareness also allows desert dwellers to consider what strategies might minimise the problem – where in the cycle shown in Figure 15 should desert people put their efforts to change the system? In true desert style, we see that there are many points of possible intervention, and in the next section we try to highlight a few key actions that could change the future of desert Australia for the better. Such a list of actions will always be incomplete, though. Our most important message is that desert dwellers (and

those outside the desert who care about it or exert power over it) should seek to understand the desert drivers; they should use them to analyse what goes wrong in their day-to-day dealings with desert life, with a view to developing more effective and long-lasting outcomes.

10.2 Caring for Australian deserts

As we have seen, over half a million people live in desert Australia – including the most significant concentration of Aboriginal people in the country – and over $90 billion of goods and services are produced annually from these three-quarters of the continent that are deeply embedded in the national identity. Getting its future on a better footing is a vital concern for all Australians.

Having exploring a wide variety of issues from the very local to the whole of desert Australia – from its ecology and pre-history to today's services and governance – we can see how they all emerge from the desert drivers, and why a better approach to each of them depends on understanding this. Some of the following responses are strictly for desert regions, some are relevant to all remote areas; some apply in truly arid country, others are important (but perhaps attenuated) in the semi-arid margins of the desert. But the following elements emerge from this book to provide a better blueprint for the desert.

Desert ecology and management. Looking across the marvellous variety of plants and animals that make the desert their home, we have seen how they adopt particular strategies to deal with the underlying desert drivers of scarce, variable and patchy resources. Whether they become ephemerals or persistent perennials, nomadic or focused on special refuge spots, or organise themselves to help each other, the weak points of each strategy are obvious. So, the local managers of natural resources – such as pastoralists, conservationists, tourism operators, park rangers or bush food harvesters – need to understand and allow for these weak points appropriately (see Chapter 5). Understanding the strategies used by plants in the desert helps us to see the equivalent weak points in other activities that depend on variable scarce resources, such as some small businesses and specialised desert settlements.

Desert livelihoods. There are relatively few livelihood options in remote areas because of sparse populations, remote markets for most products and the costs imposed by distance. To have reliable livelihoods in the desert, people need to focus on those that have a desert competitive advantage – which are often based on natural or cultural resources that do not occur elsewhere, or which are of high value to offset the high cost of transport. These include cultural and environmental tourism, mining, sustainable precision pastoralism, bush products and cultural products such as art and music. They also include natural and cultural heritage management on behalf of society, including managing issues such as fire and biodiversity, where local management saves the nation money and provides all sorts of side benefits in health and safety (Chapters 5 and 6). Most of these sources of livelihood are variable and patchy and anyone, public or private, who invests in them needs to adopt the same strategies that desert plants and animals exhibit: build reserves, facilitate mobility, take advantage of transient good times and go quiet in lean times.

Desert businesses. Small businesses in deserts struggle because of the limited size of local markets, distance from larger markets, lack of critical mass among operators and huge problems with social uncertainty, such as staff turnover and uncertain government policy and funding regimes. There are success stories, where business sectors build critical mass through long distance (and local) partnerships between like-minded businesses, or alliances with the few large businesses (such as mining and larger tourism operators). This can be assisted through business clustering (Chapter 6). Another strategy is to apply smart desert innovation to niche products that are economically viable in desert regions but will have markets elsewhere once prices are brought down, such as remote area power supplies and desert health and education services (Chapter 8). Building a strong culture of celebrating and supporting desert innovation is vital.

Desert settlements. Because of the sparse and patchy nature of desert populations, most of their settlements have odd characteristics compared with those of more settled regions. The common interests of desert communities are usually narrower (for example, many are mining, tourism, pastoral or Aboriginal settlements) and their actual nature is unconventional (some are ephemeral like many mining towns, dispersed like most pastoral settlements, or used 'nomadically' like many Aboriginal settlements and fly-in-fly-out mining camps). A few larger service centres earn their living by servicing the smaller villages. Thus administrators must pay special attention to understanding the distinctive aspirations of different communities. The communities themselves must also play a significant role in delivering their own services to their own satisfaction (Chapter 7). Where this does not happen, there are perverse and messy misconceptions about what constitutes a 'viable' desert settlement – resulting in supply-driven services, a waste of public resources, an undermining of private initiative, and even autocratic decisions to close settlements down because they do not fit the paradigm in the mind of a distant government.

Desert governance. Governance for local services presents special problems in deserts, whether those services are power and water, natural resource management, or education and health. Sparse remote populations mean that, although accountability needs to be as local as possible to be tuned to local conditions in this vast country, some services need to be delivered at a large regional scale to make them economic. Even then, delivery organisations need to combine different types of services to be efficient (economies of scope rather than scale – see Chapter 9). Attempts to have elections and service delivery at the same scale (as is common in more settled regions) therefore generally fail. Successful case studies of regional governance are based on autonomy at the most local scale ('subsidiarity'), linked with (but separate from) flexible regional organisations to deliver the services (Chapter 9). These principles of good governance design and practice can also be found in Aboriginal society.

These are key lessons for various sectors. But the desert syndrome has other general and pervasive effects that need to be acknowledged.

Paying attention to the desert. Behind the desert syndrome shown in Figure 15 is a fundamental, and quite understandable, reality for any remote area. Because it is remote, the desert is bound to have far less attention paid to it than well-populated cities and towns. Yet desert Australia is important in the psyche and economy of urban Australia, and thus of central governments. So, from time to time, interest will turn transiently to the desert, and people will want to make changes to how it operates. Desert dwellers should be aware from historical experience that the interest soon passes, so the important thing is to use the opportunity to establish structures that are resilient to a lack of interest. The structures should allow local people to make their own decisions that do not always depend on central funding and support, which will invariably change or be withdrawn (Chapter 9).

Local knowledge and culture. The desert syndrome also highlights the special significance of local knowledge and cultural differences in remote regions, whether it is the special landscape knowledge of the pastoralist or ranger, the geological understanding of the miner or, particularly, the persistence of Aboriginal culture and desert wisdom over eons. Part of the reason local knowledge is so important is that central government is just too far away to understand and sensibly manage day-to-day activities in these remote regions. This means that letting local people have the rights and responsibilities over what they do – within whatever national guidelines and standards are deemed appropriate – is vital for the future of desert Australia. This does not mean an open slate to wreck public resources, but it does mean giving people the freedom and opportunity to work with local conditions. This is one of the structural changes towards resilience that needs to be established during a period of interest in the desert.

Capital leakage. Desert Australia has been a source of wealth to the nation for the past century and a half: principally through mining and grazing exports, through cheap labour obtained from Aboriginal people in earlier years, and in tourism today. This export of natural and social capital has only been very partially compensated by return flows of financial capital from national beneficiaries. The status of natural, social and human capital in the desert has therefore been greatly run down. As we have noted, the limited supply of capital is key to maintaining the desert syndrome; this can only be countered by funding mechanisms that are independent of the waxing and waning of interest from the centres of governance. The past cannot now be re-visited, but this situation can be corrected for the future by establishing an Outback Capital Trust Fund able to levy resources rents, which can be reinvested in desert capital, supported by changes to the handling of Commonwealth Grants Commission recurrent funding (Chapter 9).

Aboriginal disadvantage and opportunity. Although this summary has not excluded Aboriginal matters, it has not, so far, highlighted them. This arises from our belief that many of the problems of Aboriginal settlements and communities derive more from their remoteness than from cultural or racial drivers, and consequently need to be tackled primarily as part of a set of remote issues for all of desert Australia, which are distinct from the concerns of more settled regions. Managing natural resources, creating livelihoods and businesses, understanding settlements and their service needs, and improving desert governance are all issues in which the difference between remote and settled areas is actually far larger than the difference between Aboriginal or non-Aboriginal concerns within the desert.

However, there is no doubt that many of these issues do need to be tackled in somewhat different ways in an Aboriginal-dominated context – for example, culturally matched livelihoods for Aboriginal people, such as managing the land, may differ from those for non-Aboriginal people. Even then, the difference may be more between people living remotely and those living in service towns, rather than between Aboriginal and non-Aboriginal people. There are Aboriginal health, education, safety, alcohol and poverty outcomes that need special attention[224], although again this should generally be because of the disastrous state of those outcomes rather than the race of those affected. Last, there certainly are some opportunities and problems with a powerful Aboriginal flavour – the maintenance of culture and language, attachment to land and a very legitimate sense of loss as a result of past policy failures and outright racism.

Thus, we would argue that each of the issues raised above needs to be considered generally as it applies to desert communities, and then considered again to ask whether there are special actions (or special approaches to general actions) needed for Aboriginal communities. Some key issues that matter for everyone, but are particularly critical for Aboriginal desert dwellers, are listed in Table 5.

To respond to these challenges, there has to be leadership, both within the desert and without.

Desert leaders. The desert syndrome calls for a new style of desert leader: one who understands the fundamental ways in which the desert system works, including which factors are outside the direct control of desert people and need to be managed for, not railed against, such as drought. The most profound issue is that all remote communities and settlements are driven by forces that differ from those

in settled areas, and which are consequently likely to be misunderstood by service agencies steeped in assumptions based on the settled areas. To counter this, desert leaders need to create alliances, broker community consensus on aspirations and make sure these are clearly expressed to policy agencies. The alliances need to be with others in the desert, with larger organisations that have a sympathy for the desert (such as some mining companies), and with key players outside the desert. Leaders need to look for the opportunity to establish governance and funding structures that make desert regions less reliant on the largesse and unreliable attention span of central governments. To do this, they need to have plans and agreed ideas in place to push through when the opportunity arises. To achieve such agreement, desert leaders need to value human networks and relationships within the desert more explicitly than is now the case.

Desert administrators. The challenge facing government officials who live in cities, but who are charged with managing aspects of desert living is, first, to understand the desert syndrome itself. Secondly, it is to appreciate that, despite their best intentions, they will be unable to remain focused on desert problems for long, because other government priorities will soon intervene. They need to appreciate they are too far away to take an interventionist approach that is sensitive to local conditions. Consequently, they should listen sympathetically to sound proposals for making those regions less dependent on distant governance and funding. Indeed, they should actively seek to devolve as much decision-making authority to a local level as is consistent with the overall policy they are implementing. And they should also try to establish flexible, resilient local structures that can withstand neglect when the focus of federal attention moves on. Of course, they need to ensure that there are measures to guard against the nepotism and corruption that can arise in small communities, but this does not mean controlling and micro-managing every move. It just means enabling local people to make their own decisions fearlessly and openly. Where central decisions about local actions have to be taken, there needs to be a strong awareness that

Table 5: Some key actions in support of Aboriginal desert dwellers.

- Support basic investments in law and order, safety, education, health, basic infrastructure and housing, and governance for long enough that communities are able to build their own support mechanisms for these
- Support livelihood development, but recognise the particular cultural appropriateness of some, such as natural-resource-related activities
- Create career growth paths supported by literacy and numeracy, and by alliances with bigger businesses such as mining
- Provide policy stability, and commit to programs that are not changed every 2 years or less
- Allow communities to express aspirations for their settlements realistically, accepting that these may be different from settled areas
- Enable Aboriginal communities to engage with new, bottom-up regional governance systems as equal partners
- Recognise, measure, value and publicise levels of social capital
- Insist that policies targeted at Aboriginal issues are based on also dealing with the general desert aspects, with no disrespect to the Aboriginal concern but to ensure that underlying drivers are being tackled effectively
- Celebrate the desert knowledge of these original Australians, invest in the maintenance of culture and language as their source of competitive advantage, and support genuine efforts to learn from it in equal partnerships

local livelihoods may be unconventional and that networks and trust may play a larger role, compared with what is typical in cities. Administrators also need to recognise the lengthy time frames and the reserves needed to deal with variability, and to clearly acknowledge how Australia values its inland.

Across all of these issues, desert people need to act together, showing respect for their different Aboriginal and non-Aboriginal needs and aspirations, but with mutual agreement to support each other against the misunderstandings of the outside world. That world is not malevolent. On the contrary, it is usually very sympathetic, but it is forgetful and prone to misunderstand the forces it is dealing with, especially those dictated by the very nature of deserts. As Rose Kunoth-Monks says, 'What I am finding in the shared journey through the desert knowledge work is that we are able to learn side by side more'.[225] And, indeed, desert Australia has much of this learning to hand on to the outside world.

10.3 Dry times for Australia

More than twenty million people live outside the desert: in rural Australia, on her coasts or in her cities. Australians' use of the country's resources, their demand for an increasing material standard of living and now their contribution to global climate change have wrought profound changes to this once isolated continent. The great cities of Australia are already experiencing water shortages. City inhabitants are becoming a little more aware of their own resource footprints, but are increasingly detached from the rural areas where that footprint falls most heavily. Many of these rural areas are facing their own resource shortages as rainfall becomes more variable. Conventional agriculture is in crisis in Australia's most productive regions, such as the Murray–Darling Basin.

In fact, the dry part of Australia is expanding. The entire continent is now subject to some disturbing trends, which are starting to resemble the desert drivers. The climate is moving into realms hitherto unexperienced: unpredictable and out of local control (Chapter 2). After a century of boundless

bounty, many resources are becoming scarce and patchy: most obviously water and energy, but also fertile soils and space. The rural–urban divide grows ever wider, with consumers expecting to pay less and less for their food despite its impact on the rural environment. Resources, people and capital are being sucked out of the regions and into the urban centres of power. Lifestyle retreats, grey nomadism and a whole new mobile class of Australians are changing the nature of settlements and community.

Not all solutions for these trends will come from the desert, of course, but there are significant lessons to be learned from those who have already lived with their effects for a long time.

As a race, humans are not very good at thinking about and planning for variability and uncertainty – we like to think that, with a bit more study or some more data, things will become more reliable and we can take a better decision. Under climate change, the future is not going to be like that. There will be more extreme events, more wild combinations of rainfall, temperature and wind that we simply have never encountered before; more unpredictability. Even as climate models improve, there will remain huge uncertainties in how biological and physical systems will respond, and how *we* will respond ourselves. Many aspects of society have become more variable in recent years too – the volatility of the financial markets and the uncertainties over food security, as well as water supplies for our cities. All Australians have to adapt to living with more uncertainty. The good news is that this is possible – desert dwellers have done it for eons! They know the strategies and the weak points (Chapter 3). They know that you need to conserve your reserves (e.g. protect water supplies), take best advantage of the more plentiful times (build up greater food stockpiles) and places (avoid paving over the best agricultural land), keep on the move (rethink our coastal settlements?), and use local knowledge (establish better local governance systems) to make yourself and your community less vulnerable.

The impact of a widening divide between rural and urban areas could be profound, because the rural areas are the source of all the basic resources that are consumed in cities. Yet, if the cities continue to draw people and resources out of the rural areas, the time may come when there is not enough food, water, nutrients and even minerals to support the cities, especially in times of drought. Cities are already feeling the pinch with water and, alarmingly, have responded by competing with agricultural areas for their water, little realising the threat this is creating to the cities' own future food security. The solution to a water shortage is not simply to put markets in place – this only ensures the water goes to the richest, not to the most important uses. To encourage people to continue to live in rural areas, they also need independence in how they govern themselves. They need access to services and the ability to make a two-way partnership with the cities. The solutions being urgently developed for the governance of desert Australia today will be needed for all rural areas tomorrow. This includes ways of making the relationship between government and community work better, and supporting the networking needed for this (Chapters 8 and 9).

The changing nature of settlements and communities is something that governments already have to respond to in rural areas as a result of lifestyle migration – the so-called 'sea-change' and 'tree-change' movements. The result is small towns without a large enough permanent population to support the local country fire brigade or volunteers to deliver meals-on-wheels. Decision makers know that they need new models of service delivery and some of these are being developed. However imperfectly, desert Australia has already worked out ways of managing for many 'unusual' settlement types. Mobile services, recognising the importance of social capital, distinguishing the scales of accountability from those of service delivery and targeting economies of scope, not of scale, in service agencies – these are all vital parts of the solution (Chapters 7, 8 and 9). The time may come when even the knowledge of how to move whole settlements effectively will be needed on the coast as sea level rise forces the abandonment of low-lying towns. For example, it may be important to know what land tenure systems need to be put in place now to enable the move to occur less painfully in 50 years time.

However, these are just the problems – there are also opportunities!

There is the opportunity to settle our land more gently. Australia's population is projected to climb from about 20 million today to between 31 and 43 million by 2056.[226] At present, the trend is for most of these additional people to sprawl out around the existing cities, taking over the best agricultural land and creating huge future security risks for the nation in terms of energy use, food supplies and social unrest. The lifestyle movements, the huge mobilisation of grey nomads and the general mobility of people today offer a great opportunity to disperse the impacts of people on the fragile Australian continent and, at the same time, to create entirely new livelihoods and business opportunities. Telecommuting, telebanking, telemedicine, telestockbroking and telemarketing all permit people to live rurally at low density while still accessing vital services. Rural and desert Australia can grasp the opportunity to create a new settlement pattern that brings people closer to their environment but still connects them to the global information stream and world markets (Chapter 6).

Through developments such as 'precision pastoralism', there is the opportunity to create a high tech, sustainable grazing industry supplying vitally needed protein to the world's expanding metropolises. As climate change bites into global food production[227] – yet increasing affluence continues to drive the rising demand for meat – it will become less and less economic to waste grain production from prime farming land on inefficiently feeding cattle for beef. Perversely, as long as it is done sustainably, it will be the dry marginal lands of the world that will be the most efficient place to grow beef without competing with grain foods and biofuels. A resurgent grazing industry in many rural and remote regions of Australia will have the opportunity to pioneer a new, high technology, global precision pastoral movement as demand for premium 'green' meat rises. Done badly, this could be disastrous for the land, but the pressures to do it will be there. Australia could lead the world by doing it well: taking advantage of new remote technologies and a deep understanding of managing for climatic variability (Chapter 5).

This is really just part of an even greater opportunity – to create a major new export market in smart dryland technologies, products, practices and knowledge more generally. Australia has always contributed ideas in dryland agriculture to the world – from drought and disease resistant Federation wheat and the self-propelled rotary hoe in times gone by, to buffalo fly traps and seasonal climate forecasting more recently.[228] Today, there is a new opportunity to re-focus on a growing international need that could, happily, be met at the same time as satisfying local priorities. In addition to dryland natural resource management, the nation should market its expertise on mobile settlements, on servicing mobile inhabitants in fixed settlements and managing sparse settlements generally to the world – as these become increasingly relevant to communities around the world. As Norway did with its oil and gas wealth before it ran out, it would be wise to take the opportunity of any resources booms to creating new knowledge-based industries such as these. A national approach to bringing together all the knowledge and innovation about managing uncertainty amid dry times online, to being recognised as the world's expert on desert knowledge, and to supporting businesses applying this knowledge, would have immense influence on the international stage.

The greatest opportunity of all is a more subtle one. The Australian psyche is built on conquering the bush: the battling settler fighting against the unfamiliar antipodean forces of nature to clear his block and break it to the plough. The achievements of those battlers were great, but their time has passed and so has this metaphor for the national psyche. Today, Australians need to tell new tales about themselves, and the greatest of these is to understand deeply the attributes of the land in which they live, with all its scarcities, unpredictability and vast distances. The opportunity is to discover that fusion of Aboriginal wisdom and modern science and weave them into the new stories that can give children pride in a sustainable future for the great

southern land, and the new songlines that can guide them on that journey.

10.4 Desert living round the world

The future is a hotter, hungrier and drier world. Today, 2 billion people inhabit the drylands of the world, half of them dependent on natural resources for their livelihoods. A few of these regions will benefit from global warming, at least for a while; they will move into cropping as the world's food supply is put under strain. But the majority of the world's great dryland belts will get drier and more desertified. Where droughts kill cows and sheep in Australia, they kill women and children in Darfur and Timbuktu. The once mighty Lake Chad in West Africa has already shrunk to a tenth of its size, yet must still support 30 million people. Dryland regions in the developing world, with their soaring populations, are already the fountainhead of unrest – even terrorism – as people are driven by lack of resources into conflict and refugee camps. Unattended, these regions will become an ever more dangerous source of international instability as global change continues.

The drylands of the world, both developing and developed, suffer the same syndrome (Figure 15) as Australia's deserts (Box 4). Some have much greater population densities, but are invariably more lightly populated than neighbouring more productive lands. Some drylands contain centres of political power – even capital cities in countries dominated by drylands, such as Saudi Arabia or Mali – but the surrounding rural hinterlands are still distant from power, and most desert countries are themselves distant from centres of world power.

So, notwithstanding some notable exceptions, the drylands of the world face the issues of desert Australia on a global scale. And at that global scale they need to band together, create critical mass and write their own futures, just as desert Australia needs to do. They face key issues to which desert knowledge can contribute.

Drylands provide, and will continue to provide, an essential part of the world's food supply, especially for meat that does not compete with staple crops. But, to do this into the future, they must find sustainable paths to grazing that developments such as precision pastoralism can both contribute to and learn from. However, land degradation in the world's drylands will continue with disastrous impacts on both global climate and local stability unless there is a serious effort to integrate the principles of desert living more effectively into the workings of the UN's Convention to Combat Desertification.[229] Drylands are a major source of potential conflict, refugees and government failure, which will extend out to their neighbours, and ultimately to the world, unless this can be stabilised with wise approaches to the desert syndrome.

Just as Australia's deserts hold lessons for urban Australia, so the world's drylands, with their deep experience and knowledge of scarcity and unpredictability, have many things to teach the unsustainable megacities of the future. For example, the drylands are in fact major exporters of embedded water to wetter areas (Chapter 8), and tend to be more efficient in their use of this scarce resource. They offer principles and insights that can help stretch out supplies elsewhere. The drylands also harbour many of the world's dying human cultures, especially nomadism, which date back to a time when humans both understood and connected with their environment. That time has gone, but the surviving wisdom is of profound value still.

For, outside the drylands, global communities more generally need to cope better with increasing uncertainty and scarcity. If the world does not adopt some desert strategies for dealing with uncertainty and scarcity, the global food system, for example, will become brittle and subject to catastrophic failure, with severe consequences for the whole of humanity. The foreshocks of this may already have emerged in the food crisis of 2007–08 when stockpiled reserves had become too small.

As we have seen, 2007 was also memorable as the first year in history in which more humans lived in cities than outside them. It seems inevitable that this trend will continue and that the gulf will grow between city consumers and the rural people who produce the food, fibre, timber, minerals and water that they need. Unless we start to think along desert lines for linking them together and providing the devolved governance systems that are needed, water wars and famine fights will erupt between city and country around the planet.

Many of the world's great civilisations were enriched by nomads bringing concepts and cultures in from the desert. The way of life, the scale of populations and the technological context has moved on, but there remains a fundamental human wisdom from the souls of those desert nomads, attuned to the unpredictable cycles of nature as they were, that needs to be remobilised as the world itself moves into dry and uncertain times.

EPILOGUE

On a beach in Lombok, a campfire blazes up. A small group of people gathered around it turn wistful eyes to the shadowy forest that has sheltered them for generations, for they know that soon they will see it no more. They are about to embark on a remarkable voyage to a new, unseen land that lies beyond the horizon. With their children, food, water, tools and weapons they will soon board several fragile bamboo rafts and sail into the sunrise. Most know they will never return. The older women and men guess that the process of adapting to their new homeland will test all their wisdom and powers of survival.

We do not know quite when this momentous event occurred. Perhaps 70 000 years ago, or 100 000 – maybe even longer. We do not know whether adventurous individuals had already scouted the new landmass of Australasia across the intervening 100 kilometres of sea, or whether a group driven by some desperate necessity abandoned their home and took to the sea. Either way, they found by good fortune a new dwelling beyond the horizon on the western shores of Papua or the Australian Kimberley. Their original landing site now lies 35 fathoms deep, at the edge of the continental shelf. Nonetheless, for a people whose distant relatives may hardly have even reached Europe yet, it was an extraordinary feat of courage, enterprise, organisation and technical skill.

Humans came to Australia only in the most recent moments of geological time. Unlike the native plants and animals, they had no evolutionary experience of living through the vast episodes of drought, fire, flood and erosion that had shaped the land and its life forms in the preceding millions of years. Fanning out into the new continent their impact was profound and dramatic. The country's ancient cycle of vast wildfires was broken, to be replaced by the Aboriginal custom of burning small patches of land constantly through the year, creating a varied, self-renewing mosaic of landscapes. And when highly skilled hunters encountered a naïve fauna of giant marsupials, the results were much the same as they had been in Europe, Asia and the Americas – the rapid disappearance of the large animals and the expansion of a diverse population of smaller and more versatile species.

It took time for some degree of balance between humans and nature to emerge. But, as the biota changed, the settlers too were changing rapidly: adapting to their new homeland. Their intimate vision and understanding of the landscape became preserved for future generations in an extraordinary body of oral compositions – the dreamtime songs – many of which survive to this day despite having originated thousands of years before the stones of the Great Pyramid of Giza or Stonehenge were hauled into place.

The dreamtime songs embody the wisdom of desert dwellers over hundreds of generations, preserving the experience of times and extreme events so rare as not to be encountered in hundreds of years, as well as the day-to-day knowledge of how to live within the landscape. They are, among other things, a response to the extreme variability of the Australian climate and landscape; a survival manual for future generations to refer to and depend on.

And, while other cultures have developed similar advice, cultural rules, taboos and secret knowledge, none has endured as a living tradition as long as that of the first Australians. Indeed, the dreaming songs in their way form an account of how these original immigrants became true Australians, finally in tune with the nature of the land that enfolded them, with its vast variability and evanescent resources.

For those arriving from Europe over the past 250 years, the journey to a shared Australian identity with those already inhabiting the continent is still unfinished. Dutch sailors who were shipwrecked or marooned on the West Australian coast in the seventeenth century do not seem to have survived – perhaps an early hint that European ways and understanding were ill-adapted to the landscape and conditions presented by the new continent. The predictable climate of Europe, the essential fertility and stability of its lands, and the settled way of life these engendered had not prepared the newcomers for the uncertainty of Australia. It had not taught them the essential survival strategies of the desert – find ways to endure through thick and thin, or else move in space or time.

The arrival of the first settlers had caused great impacts, but the effect of the second wave of arrivals has been far more intense. Drought was a regular visitor to the colony when Charles Darwin passed through in the scorching summer of 1836 and, a few years later, the explorer Paul Strzelecki observed the deadly synergy of drought and human activity, commenting on how the soil was parched by the clearing of native vegetation 'under the innumerable flocks and axes which the settlers had introduced'. Almost two centuries on, Australians are slowly learning to live with drought – at the dawning of an era when its impact is expected to increase greatly across Australia and half the world. This time we cannot take 10 000 years to come to know the continent, because the changes are coming upon us much faster.

Like the river red gum, the red kangaroo, the burrowing frog or the paper daisy, any creature that aspires to live for long in Australia must adapt its ways to the episodic character of the continent, its thin soils, sparse nutrients, variable water and climate – or perish. Of all Australians, those best adapted to the natural rhythms of the continent are those who live in closest company with its greatest variability: the inhabitants of the deserts. Knowledge from the deserts is vital to the process of becoming true Australians, all the more so faced with future climate change. We must learn from those who have gone before, and those who are innovating now.

For the Australian deserts, for Australia, for the world: desert solutions for desert problems – the dryland wisdom for the dry times to come.

REFERENCES

ABS (2001) 'ABS views on remoteness'. Australian Bureau of Statistics, Information Paper 1244.0, Canberra. <http://www.abs.gov.au/AUSSTATS/abs@.nsf/DetailsPage/1244.02001?OpenDocument> (accessed March 2009).

ABS (2008) 'Population projections, Australia, 2006 to 2101'. Australian Bureau of Statistics, Paper 3222.0, Canberra. <www.abs.gov.au/ausstats/abs@.nsf/mf/3222.0> (accessed March 2009).

Adepoyibi C (2006) Problems of Indigenous community councils in north Australia. *Australian Journal of Public Administration* **65**, 17–28.

AIATSIS/FATSIL (2005) 'National Indigenous Languages Survey Report 2005'. Department of Communications, Information Technology and the Arts, Canberra, Australia.

Allan RJ, Beard GS, Close A, Herczeg AL, Jones PD and Simpson HJ (1996) 'Mean sea level pressure indices of the El Niño–Southern Oscillation: Relevance to stream discharge in south-eastern Australia. CSIRO Division of Water Resources'. Report 96/1, Canberra, Australia.

Altman JC (1987) *Hunter Gatherers Today: An Aboriginal Economy in North Australia*. Australian Institute of Aboriginal Studies, Canberra.

Amiran DHK (1973) Problems and implications in the development of arid lands. In *Coastal Deserts: Their Natural and Human Environments*. (Eds DHK Amiran and AW Wilson) pp. 25–32. The University of Arizona Press, Tuscon, Arizona.

Anderson A (2002) Speech to Indigenous Governance Conference. In 'Indigenous Governance Conference'. Reconciliation Australia: Canberra. <http://www.reconciliation.org.au/home/media/speeches> (accessed March 2009).

Anderson A (2004) Women's rights and culture: an indigenous woman's perspective on the removal of traditional marriage as a defence under Northern Territory law. *Indigenous Law Bulletin* **2004**, Article 31 <http://www.austlii.edu.au/au/journals/ILB/2004/31.html> (accessed March 2009).

Andreasyan K and Hoy WE (2009) Patterns of mortality in Indigenous adults in the Northern Territory, 1998–2003: are people living in more remote areas worse off? *Medical Journal of Australia* **190**, 307–311.

Bastin G and the ACRIS Management Committee (2008) *Rangelands 2008 – Taking the Pulse*. National Land and Water Resources Audit, Canberra.

Biograze (2000) *Biograze: Waterpoints and Wildlife*. CSIRO, Alice Springs.

Botterill L and Fisher M (Eds) (2003) *Beyond Drought in Australia: People, Policy and Perspectives*. CSIRO Publishing, Melbourne, Australia.

Brown D, Taylor J and Bell M (2008) The demography of desert Australia. *The Rangeland Journal* **30**, 29–43.

Bureau of Infrastructure Transport and Regional Economics (2008) *About Australia's regions – June 2008*. Department of Infrastructure, Transport, Regional Development and Local Government, Canberra.

Burke EJ, Brown SJ and Christidis N (2006) Modeling the recent evolution of global drought and projections for the twenty-first century with the Hadley Centre climate model. *Journal of Hydrometeorology* **7**, 1113–1125.

Campbell D, Stafford Smith M, Davies J, Kuipers P, Wakerman J and McGregor M (2008) Responding to the health impacts of climate change in the Australia desert. *Rural and Remote Health* [online], article 1008 <http://www.rrh.org.au/articles/showarticlenew.asp?ArticleID=1008> (accessed March 2009).

Campbell L (2006) *Darby: One Hundred Years of Life in a Changing Culture*. Australian Broadcasting Corporation, Sydney.

Carter JO, Hall WB, Brook KD, McKeon GM, Day KA and Paull CJ (2000) AussieGRASS: Australian grassland and rangeland assessment by spatial simulation. In *Applications of Seasonal Climate Forecasting in Agricultural and Natural Ecosystems – The Australian Experience*. (Eds G Hammer, N Nicholls and C Mitchell) pp. 329–349. Kluwer Academic Press, Netherlands.

Chatwin B (1988) *The Songlines*. Penguin, New York.

Davidson R (2006) No fixed address: nomads and the fate of the planet. *Quarterly Essay* **24**, 1–53.

Davies J, White J, Wright A, Maru Y and LaFlamme M (2008) Applying the sustainable livelihoods approach in Australian desert Aboriginal development. *The Rangeland Journal* **30**, 55–65.

Davies PM, Bunn SE and Balcombe F (2003) 'Importance of flood flows to the productivity of dryland rivers and their floodplains: final report'. Environment Australia, Canberra. <http://www.environment.gov.au/water/publications/environmental/rivers/nrhp/flood/index.html> (accessed March 2009).

Dekker SC, Rietkerk M and Bierkens MFP (2007) Coupling microscale vegetation-soil water and macroscale vegetation-precipitation feedbacks in semiarid ecosystems. *Global Change Biology* **13**, 671–678.

Desert Knowledge Australia (2006) 'Linked business networks project: final report to AusIndustry'. Desert Knowledge Australia Alice Springs. <http://www.desertknowledge.com.au/dka/index.cfm?attributes.fuseaction=bn_pilot> (accessed March 2009).

Desert Knowledge Australia (2008) 'remoteFOCUS: revitalising remote Australia'. Desert Knowledge Australia, Alice Springs. <http://www.desertknowledge.com.au/dka/index.cfm?fuseaction=remoteFocus> (accessed March 2009).

Dillon MC and Westbury ND (2007) *Beyond Humbug: Transforming Government Engagement with Indigenous Australia*. Seaview Press West Lakes, South Australia.

Dollery BE and Fleming E (2005) 'A conceptual note on scale economies, size economies and scope economies in Australian local government'. University of New England, School of Economics, Working Paper Series in Economics 2005-6, Armidale, Australia. <http://www.une.edu.au/febl/EconStud/wps.htm> (accessed March 2009).

Drought Policy Review Task Force (1990) *Managing for Drought*. Australian Government Publishing Service, Canberra.

Edwards GP and Allan GE (Eds) (2009) 'Desert Fire: fire and regional land management in the arid landscapes of Australia'. Desert Knowledge CRC Report 37, Alice Springs. <http://www.desertknowledgecrc.com.au/publications/research.html> (accessed July 2009).

Edwards GP, Allan GE, Brock C, Duguid A, Gabrys K and Vaarzon-Morel P (2008) Fire and its management in central Australia. *The Rangeland Journal* **30**, 109–121.

Ellis S, Kanowski P and Whelan R (2004) 'National inquiry on bushfire mitigation and management'. Commonwealth of Australia, Canberra. <http://www.coagbushfireenquiry.gov.au/findings.htm> (accessed March 2009).

Esteban J and Fairen V (2006) Self-organized formation of banded vegetation patterns in semi-arid regions: a model. *Ecological Complexity* **3**, 109–118.

Evans L, Scott H, Muir K and Briscoe J (2009) Effective intellectual property protection of traditional knowledge of plants and their uses: an example from Australia. *GeoJournal* [online first: doi 10.1007/s10708-008-9229-6].

Farrelly M (2005) Regionalisation of environmental management: a case study of the Natural Heritage Trust, South Australia. *Australian Geographical Studies* **43**, 393–405.

Ferguson J (2008) Dynamic deserts and future research. In 'Desert Knowledge Symposium 2008'. Desert Knowledge CRC/Desert Knowledge Australia, Alice Springs. <http://www.desertknowledgecrc.com.au/publications/downloads/DKCRC_Dynamic-deserts-and-future-research.pdf> (accessed March 2009).

Flannery T (2007) Trajectory of human evolution in Australia: 10,000 ybp–present. In *Sustainability or Collapse: An Integrated History and Future of People on Earth* (Eds R Costanza, LJ Graumlich and W Steffen) pp. 89–94. MIT Press (with Dahlem University Press), Cambridge, Mass.

Flores J and Jurado E (2003) Are nurse-protégé interactions more common among plants from arid environments? *Journal of Vegetation Science* **14**, 911–916.

Folland CK, Renwick JA, Salinger MJ and Mullan AB (2002) Relative influences of the Interdecadal Pacific Oscillation and ENSO on the South Pacific Convergence Zone. *Geophysical Research Letters* **29**, 10.1029.

Foran BD and Stafford Smith DM (1991) Risk, biology and drought management strategies for cattle stations in Central Australia. *Journal of Environmental Management* **33**, 17–33.

Giles E (1889) *Australia Twice Traversed*. Facsimile [1986] of original edition, Doubleday Australia, Sydney.

Gilfillan SL (2001) An ecological study of a population of *Pseudantechinus macdonnellensis* (Marsupialia : Dasyuridae) in central Australia. II. Population dynamics and movements. *Wildlife Research* **28**, 481–492.

Gore A (2007) *The Assault on Reason*. Penguin Press, New York.

Grey-Gardner R (2008) Implementing risk management for water supplies: a catalyst and incentive for change. *The Rangeland Journal* **30**, 149–156.

Grove RH (2007) Revolutionary weather: the climatic and economic crisis of 1788–1795 and the discovery of El Niño. In *Sustainability or Collapse: An Integrated History and Future of People on Earth*. (Eds R Costanza, LJ Graumlich and W Steffen) pp. 151–168. MIT Press (with Dahlem University Press), Cambridge, Mass.

Gullan PJ and Cockburn A (1986) Sexual dichronism and intersexual phoresy in gall-forming coccoids. *Oecologia* **68**, 632–634.

Gunya Australia (2007) 'Indigenous Economic Development Scheme: a solution to create employment opportunities within Indigenous communities'. Gunya Australia, Sydney. <http://gunya.com.au/acrobat/gunya_discussion_paper_august_07.pdf> (accessed March 2009).

Hennessy K, Fawcett R, Kirono D, Mpelasoka F, Jones D, Bathols J, Whetton P, Stafford Smith M, Howden M, Mitchell C and Plummer N (2008) 'An assessment of the impact of climate change on the nature and frequency of exceptional climatic events: report to Australian Government'. Bureau of Meteorology and CSIRO, Canberra. <http://www.daff.gov.au/agriculture-food/drought/national_review_of_drought_policy/climatic_assessment> (accessed March 2009).

Hennessy K, Fitzharris B, Bates BC, Harvey N, Howden SM, Hughes L, Salinger J and Warrick R (2007) Australia and New Zealand. In *Climate Change 2007: Impacts, Adaptation and Vulnerability. Contribution of Working Group II to the Fourth Assessment Report of the Intergovernmental Panel on Climate Change*. (Eds ML Parry, OF Canziani, JP Palutikof, PJ van der Linden and CE Hanson) pp. 507–540. Cambridge University Press, Cambridge, UK.

Hercus L (1985) Leaving the Simpson Desert. *Aboriginal History* **9**, 22–43.

HilleRisLambers R, Rietkerk M, van den Bosch F, Prins HHT and de Kroon H (2001) Vegetation pattern formation in semi-arid grazing systems. *Ecology* **82**, 50–61.

Holmes J (1997) Diversity and change in Australia's rangeland regions: translating resource values into regional benefits. *The Rangeland Journal* **19**, 3–25.

Hunt J and Smith D (2007) 'Indigenous Community Governance Project: year two research findings'. CAEPR, ANU, Canberra. <http://www.anu.edu.au/caepr/ICGP_publications.php#resreports> (accessed Aug 2007).

James CD (1994) Spatial and temporal variation in structure of a diverse lizard assemblage in arid Australia. In *Lizard Ecology: Historical and Experimental Perspectives*. (Eds LJ Vitt and ER Pianka) pp. 287–317. Princeton University Press, Princeton, New Jersey, USA.

James D (2005) Kinship with country – acts of translation in the cross-cultural performance space. A case study on the Anangu Pitjantjatjara Lands of Central Australia. PhD Thesis, Australian National University, Canberra.

Johnson, K (1992) *The AUSMAP Atlas of Australia*. Cambridge University Press, Cambridge, UK.

Johnston PW, Tannock PR and Beale IF (1996) Objective 'safe' grazing capacities for south-west Queensland, Australia: Model application and evaluation. *The Rangeland Journal* **18**, 259–269.

Keen I (2004) *Aboriginal Economy and Society: Australia at the Threshold of Colonisation*. Oxford University Press, South Melbourne.

Ker Conway J (1989) *The Road from Coorain*. Random House, Sydney.

Kerwin D (2006) *Aboriginal Dreaming Tracks or Trading Paths: The Common Ways*. Griffith University, Brisbane.

Kimber RG (1986) *Man from Arltunga*. Hesperian Press, Perth.

Kunoth-Monks R (2006) Land and culture: necessary but not sufficient for the future. Identity in the 21st Century. In 'Desert Knowledge Symposium'. (Desert Knowledge Australia and Desert Knowledge CRC, Alice Springs).

LaFlamme M (2007) Developing a shared model for sustainable Aboriginal livelihoods in natural-cultural resource management. In *Land, Water and Environmental Management: Integrated Systems for Sustainability. MODSIM 2007 International Congress on Modelling and Simulation*. (Eds L Oxley and D Kulasiri) pp. 288–294. Modelling and Simulation Society of Australia and New Zealand, Christchurch, New Zealand.

Lane MB, Cheers B and Morrison TH (2005) Regionalised natural resource management in South Australia: prospects and challenges of the new regime. *South Australian Geographical Journal* **104**, 11–25.

Lane MB, McDonald GT and Morrison T (2004) Decentralisation and environmental management in Australia: a comment on the prescriptions of the Wentworth Group. *Australian Geographical Studies* **42**, 102–114.

Lange RT, Nicolson AD and Nicolson DA (1984) Vegetation management of chenopod rangelands in South Australia. *Australian Rangelands Journal* **6**, 46–54.

Langton M (2007) Trapped in the Aboriginal reality show. *Griffith REVIEW* **19**, 1–19.

Larson S (2007) 'An overview of the natural resources management arrangements in the Lake Eyre Basin'. Desert Knowledge CRC and CSIRO Sustainable Ecosystems, Alice Springs and Townsville. <http://www.csiro.au/resources/NRM-Lake-Eyre.html> (accessed March 2009).

Latz P (2007) *The Flaming Desert: Arid Australia – A Fire Shaped Landscape*. Peter Latz, Alice Springs.

Ludwig JA, Tongway DJ and Marsden SG (1999) Stripes, strands or stipples: modelling the influence of three landscape banding patterns on resource capture and productivity in semi-arid woodlands, Australia. *Catena* **37**, 257–273.

Ludwig JA, Wilcox BP, Breshears DD, Tongway DJ and Imeson AC (2005) Vegetation patches and runoff-erosion as interacting ecohydrological processes in semiarid landscapes. *Ecology* **86**, 288–297.

Maclean K (2009) Re-conceptualising desert landscapes: unpacking historical narratives and contemporary realities for sustainable livelihood development in central Australia. *GeoJournal* [online first, doi:10.1007/s10708-008-9234-9].

Marshall GR (2008) Nesting, subsidiarity, and community-based environmental governance beyond the local level. *International Journal of the Commons* **2**, 75–97.

Maru YT and Chewings VH (2008) How can we identify socio-regions in the rangelands of Australia? *The Rangeland Journal* **30**, 45–53.

Maru YT, McAllister RRJ and Stafford Smith M (2007) Modelling community interactions and social capital dynamics: The case of regional and rural communities of Australia. *Agricultural Systems* **92**, 179–200.

Mayes JC (2004) London's wettest summer and wettest year – 1903. *Weather* 59, 274–278.

McAllister RRJ, Cheers B, Darbas T, Davies J, Richards C, Robinson CJ, Ashley M, Fernando D and Maru YT (2008) Social networks in arid Australia: a review of concepts and evidence. *The Rangeland Journal* 30, 167–176.

McAlpine CA, Syktus J, Deo RC, Lawrence PJ, McGowan HA, Watterson IG and Phinn SR (2007) Modeling the impact of historical land cover change on Australia's regional climate. *Geophysical Research Letters* 34(L22711), 6 pp.

McBryde I (2000) Travellers in storied landscapes: a case study in exchanges and heritage. *Aboriginal History* 24, 152–174.

McCormack J (1989) Stocking rates and land care: the Merino Downs case. *Australian Rangelands Society Newsletter* 1989, 12–14.

McDonald R (1991) *Winning the Gascoyne*. Hesperian Press, Perth.

McGowan B (2002) *Australian Ghost Towns*. Lothian Books, Melbourne.

McKeon G, Hall W, Henry B, Stone G and Watson I (Eds) (2004) *Pasture Degradation and Recovery in Australia's Rangelands: Learning from History*. Queensland Department of Natural Resources, Mines and Energy, Brisbane.

MEA (2005) *Millennium Ecosystem Assessment – Ecosystems and Human Well-being: Desertification Synthesis*. World Resources Institute, Washington DC.

Measham TG, Robinson C, Richards C, Larson S, Stafford Smith M and Smith T (2007) 'Tools for successful NRM in the Lake Eyre Basin: achieving effective engagement'. Desert Knowledge CRC, Alice Springs. <http://www.csiro.au/resources/NRM-Lake-Eyre-Community-Engagement.html> (accessed March 2009).

Memmott P (2005) Tangkic Orders of Time: an anthropological approach to time study. *Journal of Media Arts Culture* 2, 1–10.

Memmott P and Moran M (2001) 'Indigenous Settlements of Australia'. Technical paper. Department of the Environment and Heritage, Canberra. <http://www.environment.gov.au/soe/2001/publications/technical/indigenous/index.html> (accessed March 2009).

Middleton NJ and Thomas DSG (1997) *World Atlas of Desertification*. 2nd edn. U.N. Environment Programme, Edward Arnold, New York.

Mildrexler DJ, Zhao M and Running SW (2006) Where are the hottest spots on Earth? *Eos* 87, 461, 467.

Mitchell J, Pearce R, Stephens M, Taylor J and Warchivker I (2005) *Indigenous Populations and Resource Flows in Central Australia: A Social and Economic Baseline Profile*. Centre for Remote Health, Alice Springs.

Molden D (Ed.) (2007) *Water for Food, Water for Life: A Comprehensive Assessment of Water Management in Agriculture (Summary)*. Earthscan and International Water Management Institute, London and Colombo.

Monbiot G (2006) *Heat: How to Stop the Planet Burning*. Penguin, London.

Moran M (2008) Demand responsive services: towards an analytical framework for administrative practice in Indigenous settlements. *Australian Journal of Public Administration* 67, 186-199.

Moran M and Elvin R (2009) Coping with complexity: adaptive governance in desert Australia. *GeoJournal* [online first, doi:10.1007/s10708-008-9240-y].

Moran M, Wright A, Renehan P, Szava A, Beard N and Rich E (2007) 'The transformation of assets for sustainable livelihoods in a remote Aboriginal settlement'. Centre for Appropriate Technology and Desert Knowledge Cooperative Research Centre Report 28, Alice Springs. <http://www.desertknowledgecrc.com.au/publications/research.html> (accessed March 2009).

Morphy H (2005) Indigenous Art as Economy. In *Culture, Economy and Governance in Aboriginal Australia*. (Eds D Austin-Broos and G Macdonald) pp. 19–28. Sydney University Press, Sydney.

Morrissey JG and O'Connor REY (1989) 28 years of station management, 'a fair use and a fair go'. *W.A. Dept. of Agriculture Pastoral Memo (Carnarvon)* **13**, 2–9.

Morton SR, Masters P and Hobbs TJ (1993) Estimates of abundance of burrowing frogs in spinifex grasslands of the Tanami Desert, Northern Territory. *The Beagle* **10**, 67–70.

Morton SR, Stafford Smith DM, Friedel MH, Griffin GF and Pickup G (1995) The stewardship of arid Australia: ecology and landscape management. *Journal of Environmental Management* **43**, 195–217.

NEGKLCD (1993) *Mulga, Merinos and Managers: A Handbook of Recommended Pastoral Management Practices*. North Eastern Goldfields and Kalgoorlie Land Conservation Districts, Kalgoorlie, WA.

Nelson R, Howden M and Stafford Smith M (2008) Using adaptive governance to rethink the way science supports Australian drought policy. *Environmental Science and Policy* **11**, 588–601.

Ngaanyatjarra Council (2003) *Doing Business with Government*. Ngaanyatjarra Council, Alice Springs.

Nicholls N and Wong KK (1990) Dependence of rainfall variability on mean rainfall, latitude, and the Southern Oscillation. *Journal of Climate* **3**, 163–170.

Orians GH and Milewski AV (2007) Ecology of Australia: the effects of nutrient-poor soils and intense fires. *Biological Reviews* **82**, 393–423.

Ostrom E (1999) Coping with tragedies of the commons. *Annual Review of Political Science* **2**, 493–535.

Paton S, Woods M, Curtis A and McDonald G (2005) 'Regional NRM – a social and environmental experiment'. Land & Water Australia, Canberra.

Pawu-Kurlpurlurnu WJ, Holmes M and Box L (2008) 'Ngurra-kurlu: A way of working with Warlpiri people'. Desert Knowledge Cooperative Research Centre Report No. 41, Alice Springs.

Pickup G (1985) The erosion cell – a geomorphic approach to landscape classification in range assessment. *Australian Rangeland Journal* **7**, 114–121.

Pickup G (1991) Event frequency and landscape stability on the floodplain systems of arid central Australia. *Quaternary Science Reviews* **10**, 463–473.

Pittock AB (2008) Climate change: desert challenge and opportunity. In 'Desert Knowledge Symposium 2008'. Desert Knowledge CRC/Desert Knowledge Australia, Alice Springs. <http://www.desertknowledgecrc.com.au/publications/downloads/DKCRC_Climate-Change-Desert-Challenge-and-Opportunity.pps> (accessed March 2009).

Productivity Commission (2008) 'Government drought support, draft inquiry report'. Productivity Commission, Melbourne. <http://www.pc.gov.au/projects/inquiry/drought/draft> (accessed March 2009).

Purvis JR (1986) Nurture the land: My philosophies of pastoral management in central Australia. *Australian Rangeland Journal* **8**, 110–117.

Read JL (1999) Abundance and recruitment patterns of the trilling frog (*Neobatrachus centralis*) in the Australian arid zone. *Australian Journal of Zoology* **47**, 393–404.

Reynolds JF, Stafford Smith DM, Lambin EF, Turner BL, II, Mortimore M, Batterbury SPJ, Downing TE, Dowlatabadi H, Fernandez RJ, Herrick JE, Huber-Sannwald E, Jiang H, Leemans R, Lynam T, Maestre FT, Ayarza M and Walker B (2007) Global desertification: building a science for dryland development. *Science* **316**, 847–851.

Rietkerk M, Dekker SC, de Ruiter PC and van de Koppel J (2004) Self-organized patchiness and catastrophic shifts in ecosystems. *Science* **305**, 2004–2009.

Robins L (2005) Partners in business: the Outback Spirit Supply Chain Model. In 'International Conference on Engaging Communities'. Queensland Government, Brisbane. <http://www.engagingcommunities2005.org/abstracts/Robins-Lisa-final.pdf> (accessed March 2009).

Roderick ML, Rotstayn LD, Farquhar GD and Hobbins MT (2007) On the attribution of changing pan evaporation. *Geophysical Research Letters* **34**, 6.

Rola-Rubzen MF and McGregor M (2008) Impact of the desert economy. In 'Desert Knowledge Symposium 2008'. Desert Knowledge CRC/Desert Knowledge Australia, Alice Springs. <http://www.desertknowledgecrc.com.au/publications/downloads/DKCRC_Impact-of-the-Desert-Economy.pdf> (accessed March 2009).

Rose DB (1996) *Nourishing Terrains: Australian Aboriginal Views of Landscape and Wilderness.* Australian Heritage Commission, Commonwealth of Australia, Canberra.

Roshier DA, Robertson AI, Kingsford RT and Green DG (2001) Continental-scale interactions with temporary resources may explain the paradox of large populations of desert waterbirds in Australia. *Landscape Ecology* **16**, 547–556.

Rothwell N (2007) Parallel Stories. *Weekend Australia Magazine* 5–6 May 2007, 18–19.

Rothwell N (2008) Indigenous insiders chart an end to victimhood. *The Australian,* 3 Sep 2008.

Ryan P (2004) 'Indigenous community engagement in safety and justice issues: A discussion paper. Aboriginal Law and Justice Strategy'. Northern Territory Government, Darwin. <http://ntru.aiatsis.gov.au/ifamp/research/research_frameset.html> (accessed March 2009).

Ryan P and Antoun J (2001) 'A model for social change: The Northern Territory's Aboriginal Law and Justice Strategy 1995–2001'. NT Department of Community Development, Sport, Culture and the Arts, Darwin.

Safriel U, Adeel Z, Niemeijer D, Puigdefabregas J, White R, Lal R, Winslow M, Ziedler J, Prince S, Archer E and King C (2005) Dryland systems. In *Millennium Ecosystem Assessment: Ecosystems and Human Well-being: Current State and Trends: Findings of the Condition and Trends Working Group.* (Eds RM Hassan, R Scholes and N Ash) pp. 623–662. Island Press, Washington DC.

Sanders W and Holcombe S (2008) Sustainable governance for small desert settlements: learning from the multi-settlement regionalism of Anmatjere Community Government Council. *The Rangeland Journal* **30**, 137–147.

Schulke B (2007) 'Beefing up grazing land: presentation to Beef-Up Forum, Biloela, 15 Feb 2007'. Meat and Livestock Australia, Brisbane. <http://www.mla.com.au/TopicHierarchy/IndustryPrograms/NorthernBeef/BeefUp+forums/Biloela/Grazing+land+management.htm> (accessed March 2009).

Seemann K, Parnell M, McFallan S and Tucker S (2008) 'Housing for livelihoods: The lifecycle of housing and infrastructure through a whole-of-system approach in remote Aboriginal settlements'. Desert Knowledge Cooperative Research Centre Report No. 29: Alice Springs.

Shaw B (1995) *Our Heart is Our Land: Aboriginal Reminiscences from the Western Lake Eyre Basin.* Aboriginal Studies Press, Canberra.

Slonosky VC (2002) Wet winters, dry summers? Three centuries of precipitation data from Paris. *Geophysical Research Letters* **29**, 1.1–1.4.

Smajgl A, Leitch A and Lynam T (Eds) (2009) 'Outback Institutions: An application of the Institutional Analysis and Development (IAD) framework to four case studies in Australia's outback'. Desert Knowledge CRC Report 31, Alice Springs. <http://www.desertknowledgecrc.com.au/publications/research.html> (accessed July 2009).

Smith A (2006) Indigenous development – without community, without commerce. *Australian Policy Online [Australian Review of Public Affairs]* <http://www.australianreview.net/digest/2006/09/smith.html> (accessed March 2009).

Smith D (2007) From COAG to coercion: a story of governance failure, success and opportunity in Australian Indigenous Affairs. In 'ANZSOG conference, Governing Through Collaboration: Managing Better Through Others'. Australia and New Zealand School of Government, Canberra.

Smith MA and Hesse P (Eds) (2005) *23 Degrees South: Archaeology and Environmental History of the Southern Deserts*. National Museum of Australia Press, Canberra.

Smith T (2006) Welfare, enterprise, and Aboriginal community: the case of the Western Australian Kimberley region, 1968–96. *Australian Economic History Review* **46**, 242–267.

Spencer B (Ed.) (1896a) *Report on the Work of the Horn Scientific Expedition to Central Australia, Part I – Introduction, Narrative and Summary*. [1994 Facsimile: Corkwood Press, Bundaberg] Melville, Mullen and Slade, Melbourne.

Spencer B (Ed.) (1896b) *Report on the Work of the Horn Scientific Expedition to Central Australia, Part II – Zoology*. [1994 Facsimile: Corkwood Press, Bundaberg] Melville, Mullen and Slade, Melbourne.

Spencer WB and Gillen FJ (1899) *The Native Tribes of Central Australia*. Macmillan, London.

Stafford Smith DM and Morton SR (1990) A framework for the ecology of arid Australia. *Journal of Arid Environments* **18**, 255–278.

Stafford Smith DM, McKeon GM, Watson IW, Henry BK, Stone GS, Hall WB and Howden SM (2007) Land Change Science Special Feature: Learning from episodes of degradation and recovery in variable Australian rangelands. *Proceedings of the National Academy of Sciences* **104**, 20690–20695.

Stafford Smith M (2003) Linking environments, decision-making and policy in handling climatic variability. In *Beyond Drought: People, Policy and Perspectives*. (Eds L Botterill and M Fisher) pp. 131–151. CSIRO Publishing, Melbourne.

Stafford Smith M (2008) The 'desert syndrome' – causally-linked factors that characterise outback Australia. *The Rangeland Journal* **30**, 3–14.

Stafford Smith M and McAllister RRJ (2008) Managing arid zone natural resources in Australia for spatial and temporal variability – an approach from first principles. *The Rangeland Journal* **30**, 15–27.

Stafford Smith M, Moran M and Seemann K (2008) The 'viability' and resilience of communities and settlements in desert Australia. *The Rangeland Journal* **30**, 123–135.

Stanley O (2008) 'A survey of the ideas and literature on demand responsive services for desert settlements: An economist's viewpoint'. Desert Knowledge CRC (Working Paper 23), Alice Springs. <http://www.desertknowledgecrc.com.au/publications/workingpapers.html> (accessed March 2009).

Stanley S and De Deckker P (2002) A Holocene record of allochthonous, aeolian mineral grains in an Australian alpine lake; implications for the history of climate change in southeastern Australia. *Journal of Paleolimnology* **27**, 207–219.

Steffen W, Burbidge AA, Hughes L, Kitching R, Lindenmayer D, Musgrave W, Stafford Smith M and Werner PA (2009) *Australia's Biodiversity and Climate Change*. CSIRO Publishing, Melbourne.

Steffen W, Sanderson A, Tyson PD, Jäger J, Matson PA, Moore III B, Oldfield F, Richardson K, Schellnhuber HJ, Turner II BL and Wasson RJ (2004) *Global Change and the Earth System: A Planet Under Pressure*. Springer-Verlag, Heidelberg.

Stewart AJ, Blake DH and Ollier CD (1986) Cambrian river terraces and ridgetops in Central Australia – oldest persisting landforms. *Science* **233**, 758–761.

Strehlow TGH (1970) Geography and the totemic Landscape in Central Australia: a functional study. In *Australian Aboriginal Anthropology*. (Ed. RM Berndt) pp. 92–140. Australian Institute of Aboriginal Studies/UWA Press, Perth.

Stuart JM (1865) *Explorations in Australia: The Journals of John McDouall Stuart*. [1984 Facsimile. Hesperian Press, Carlisle, Western Australia] Saunders, Otley and Co, London.

Taylor J (2002) 'The spatial context of Indigenous service delivery'. Centre for Aboriginal Economic Policy Research, Australian National University, Canberra.

Taylor J, Ffowcs-Williams I and Crowe M (2008) Linking desert businesses: the impetus, the practicalities, the emerging pay-offs, and building on the experiences. *The Rangeland Journal* **30**, 187–195.

Taylor J and Stanley J (2005) 'The opportunity costs of the status quo in the Thamarrurr region'. Working Paper No. 28/2005, Centre for Aboriginal Economic Policy Research, Australian National University, Canberra.

Turner MK (2005) Everything comes from the land. Poster. IAD Press, Alice Springs.

Vaarzon-Morel P and Gabrys K (2009) Fire on the horizon: contemporary Aboriginal burning issues in the Tanami Desert, central Australia. *GeoJournal* [online first, doi: 10.1007/s10708-008-9235-8].

Van Kranendonk MJ, Smithies RH and Bennett VC (2007) *Earth's Oldest Rocks*. Elsevier, Amsterdam.

Verstraete MM, Scholes RJ and Stafford Smith M (2008) Climate and desertification: looking at an old problem through new lenses. *Frontiers in Ecology and the Environment* **7**, in press [eview doi: 10.1890/080119].

Wales-Smith BG (1973) Analysis of monthly rainfall totals representative of Kew, Surrey from 1697 to 1970. *Meteorological Magazine* **102**, 157–171.

Walsh F (2009) To hunt and to hold: Martu Aboriginal people's uses and knowledge of their country, with implications for co-management in Karlamilyi National Park and the Great Sandy Desert. PhD Thesis, Schools of Anthropology and Plant Biology, The University of Western Australia, Perth.

Westbrook JI, Coiera EW, Brear M, Stapleton S, Rob MI, Murphy M and Cregan P (2008) Impact of an ultrabroadband emergency department telemedicine system on the care of acutely ill patients and clinicians' work. *Medical Journal of Australia* **188**, 704–708.

Wild R and Anderson P (2007) '*Ampe Akelyernemane Meke Mekarle* "Little children are sacred": Report of the Northern Territory Board of Inquiry into the Protection of Aboriginal Children from Sexual Abuse'. NT Government, Darwin.

Wynn Jones R (2006) *Applied Palaeontology*. Cambridge University Press, Cambridge, UK.

ENDNOTES

1. *Use of the terms Aboriginal/Indigenous*: Indigenous in Australia is used to refer to Aboriginal and Torres Strait Islander people in general; no Torres Strait Islanders traditionally live in desert Australia, so we only use the term 'Indigenous' when we specifically mean to include both, or when referring to a term or organisation that is formally defined with the word Indigenous. We capitalise either term when it refers to Australian people, but a lower case 'indigenous' is occasionally used to refer to indigenous people or products more generally.

2. p. 27 in Giles (1889) Vol. 1.

3. Brown *et al.* (2008).

4. Vegetation map simplified from <http://www.desertknowledgecrc.com.au/desert_information/desertmaps.html> (accessed March 2009) with spinifex shading added from Johnson (1992, p. 58).

5. Nicholls and Wong (1990); this assertion, and those following, about the drivers of desert Australia are expanded upon with data in Stafford Smith (2008). See Chapter 3 for more details.

6. Gilfillan (2001); see also Figure 5a in Chapter 3.

7. James (1994).

8. Morton *et al.* (1993), and also Read (1999).

9. These strategies are explored in more detail in Chapter 3.

10. Chapter 2 in Ellis *et al.* (2004).

11. Mildrexler *et al.* (2006). This is a satellite measure of the integrated radiative heat of 25×25 m^2 areas of the Earth's surface, and so is not readily comparable with normal meteorological observations; however, this study reported the first comprehensive measures of the whole world's surface.

12. Oldest rocks: Van Kranendonk *et al.* (2007). Oldest continuously exposed surfaces: Stewart *et al.* (1986).

13. Davies *et al.* (2003). They summarise: 'Rates of terrestrial NPP for the Lake Eyre Basin (based on current land-use practices and average rainfall) are about 0.1 t C/ha/year. In contrast, estimated NPP for the littoral areas of the pools (the 'bath-tub ring' …) is over 50× this value. Rates of NPP in the bathtub ring are similar to primary rainforest or highly irrigated areas of the south east coast of Australia …' See also their Table 4.

14. p. 416 in Spencer (1896b).

15. The gall-forming wasp is *Cystococcus pomiformis* and the bloodwood host tree is the eucalypt *Corymbia opaca*. See <http://www.alicespringsdesertpark.com.au/kids/nature/inverte/gall.shtml> (accessed March 2009) as well as Gullan and Cockburn (1986).

16. Roshier *et al.* (2001).

17. p. 111 in Spencer (1896a).

18. Foran and Stafford Smith (1991) used a computer model to simulate how much money two neighbouring cattle properties in central Australia would have made over the past century if they had been running their current management strategies all that time. They used the same 100 years of climate data for both properties – it was hugely variable as usual. In fact, the variable strategy did better than the constant strategy after tax and subsidies were included, but probably put the land at greater risk of degradation, an issue we shall return to in Chapter 5.

19. p. 37 in Stuart (1865).

20. Rola-Rubzen and McGregor (2008).

21. <http://www.innovation.gov.au/Section/AboutDIISR/FactSheets/Pages/Australia'sExportsFactSheet.aspx> and <http://www.budget.gov.au/2008%2D09/content/bp1/html/bp1_bst10-03.htm> (accessed March 2009): from these papers gross government revenue in 2007/08 is about $319.4 billion/year, which is 25.9 per cent of gross domestic product, which is hence about $1.233 trillion.

22. Bureau of Infrastructure Transport and Regional Economics (2008).

23. E.g. see Wynn Jones (2006).

24. From *Mulga and spinifex plain*, lyrics by Neil Murray, sung by the Warumpi Band on their 1985 album *Big Name, No Blankets*. 'Tjilpi' is the respectful Luritja word for 'old man'.

25. The map is re-drawn (on a different projection) from Middleton and Thomas (1997) as Fig. 22.1 in MEA (2005); data from Table 22.1 in MEA (2005). A desert, as we use the term here, is defined as a place with an aridity index of 0.5 or less, while drylands include lands with an aridity index of up to 0.65. Areas with an index of less than 1.0 have an annual moisture deficit, meaning their average rates of evaporation exceed their rainfall. A dryland is thus a place where the potential moisture loss is at least 50 per cent more than its potential gain on average. (These terms are defined more formally in Middleton and Thomas 1997.)

26. For an account of these southern deserts, their environment, history and archaeology, see Smith and Hesse (2005).

27. Flannery (2007).

28. De Deckker, P., *pers. comm.*; see also Stanley and De Deckker (2002).

29. E.g. McAlpine *et al.* (2007).

30. E.g. Grove (2007).

31. Nicholls and Wong (1990).

32. In fact, the El Niño cycle ranges from 1 to 8 years during 1891–1994: see Allan *et al.* (1996).

33. Folland *et al.* (2002).

34. Redrawn from Figure 2 in Stafford Smith (2008), wherein some other examples for stations with different mean annual rainfalls and seasonality may also be found. In the graph, an 'event' is formally defined as one or more continuous days of rainfall. The columns represent the number of events of *at least* the size shown (with a logarithmic y-scale), thus they are cumulative towards the origin, which gives a nearly log-linear decline with increasing event size (noting that the interval is different for the first category).

35. Pickup (1991).

36. Pers. comm. Dr John Childs, Alice Springs (17 Mar 2008).

37. Stafford Smith and Morton (1990), and Stafford Smith (2008).

38. It is worth noting that recent work at the Australian National University by Mike Roderick and Graham Farquhar points out that evaporative demand should actually decrease with global warming on average, because the higher temperatures cause more evaporation and so the atmosphere should come into balance and actually appear wetter; however, this depends on water being non-limiting and may only occur close to coasts and major water bodies. In their publication (Roderick *et al.* 2007), there are in fact signs that while the evaporative demand averaged for weather stations across Australia is decreasing, it may be increasing at desert sites. Hence the assertion of increasingly harsh growing conditions in deserts made in this paragraph is probably still true.

39. <http://www.unccd.int/publicinfo/factsheets/showFS.php?number=10> (accessed March 2009).

40. Redrawn from Fig. 3a in Burke *et al.* (2006), on to the same projection as Box 2, and to highlight areas where the index is distinctly different from zero – the grey areas are distinctly drying, the black areas have become distinctly wetter and the remainder is in-between (note that the arctic and antarctic regions were not included in the original analysis). The observation of these climate changes does not in itself prove that they are human-induced, but they are broadly compatible with what climate models would suggest should have happened given known changes in the atmosphere over the twentieth century.

41. Burke *et al.* (2006): technically, their moderate, severe and extreme drought conditions are defined for each grid cell of the Earth as years in the bottom 20th, 5th or 1st percentile of the corrected Palmer Drought Index calculated for that grid cell over time; projecting forwards, then, the moderate drought conditions that *were* being experienced in each grid cell 20 per cent (for example) of the time in the past are expected to occur 50 per cent of the time by 2090 (and 5 per cent to 40, 1 to 30 per cent for the severe and extreme drought conditions, respectively).

42. Hennessy *et al.* (2008).

43. Hennessy *et al.* (2007).

44. This map is regularly updated at <http://www.bom.gov.au/cgi-bin/silo/reg/cli_chg/trendmaps.cgi> (accessed March 2009, and re-drawn and simplified from the version available at that date). The specific choice of start date for this map greatly affects the appearance of the figure. World temperature traces (e.g. see <http://data.giss.nasa.gov/gistemp/graphs/> [accessed March 2009]) and those for Australia show strong signs of a discontinuity around 1960–70, with a weaker rising trend before this since the 1900s (and a pause after World War II), and a stronger upwards trend thereafter. This change in trend is stronger in the more heavily populated and industrialised northern hemisphere (also with more land mass) than in the southern. This timing matches with inflexions in many other historical curves (e.g. see Fig. 9 in Steffen *et al.* 2004), including population, car and phone usage, globalisation, and various other factors that can be regarded as a credible basket of indicators of increasing energy use intensity. These all had direct effects on CO_2 release, but equally important indirect effects through increasing pressure on nitrogen, water, phosphorus and other cycles (many of which also show an upsurge in rates of change around this time). The Bureau of Meteorology site allows the user to readily explore alternative start dates. *All* start dates from 1900–1970 show drying in the eastern coastal strip and south-west of Australia. Some decades show this drying over a larger area of the continent. Starting in the 1960s gives a somewhat misleading view because this was a drought period over the eastern part of the continent; likewise the 1970s were particularly wet. We therefore illustrate 1950–2008, which shows drying patterns consistent with expected changes in ENSO.

45. In this work, a drought is defined as conditions below the soil moisture deficit experienced in the worst 10 per cent of years from 1974–2003 – see p. 515 in Hennessy *et al.* (2007).

46. Millennium Ecosystem Assessment: Dryland Systems: Safriel *et al.* (2005).

47. The statistics in this paragraph also come from Safriel *et al.* (2005) – some apply to deserts, some only to the wider category of drylands (i.e. including sub-humid regions, see Box 2); however, the general scale of issues highlights the significance of deserts, regardless of these details.

48. Wayne Meyer <http://www.amonline.net.au/factSheets/water_use.htm> (accessed March 2009): his figures are actually for irrigated pastures where the production system uses water more efficiently (not necessarily more appropriately) than in rangelands; but rough calculations for rangelands produce similar figures. As Meyer (*pers. comm.*) points out, the issue is not whether this use of water is right or wrong, but whether there are alternative uses that are more valuable. In irrigated regions there often are – other forms of production, environmental flows, drinking water for cities, etc.; in deserts there may not be. Here the point is simply to recognise that grazing production does use a great deal of water.

49. See <http://www.iwmi.cgiar.org/news_room/pdf/Bitter_Harvest-Deep_Focus-Sunday_Specials-Opinion.pdf> (accessed March 2009).

50. See UN Environment Program, 2006 at <http://www.unep.org/wed/2006/downloads/PDF/FactSheetWED2006_eng.pdf> (accessed March 2009).

51. See Chapter 3 in the Millennium Ecosystems Assessment's Desertification Synthesis MEA (2005).

52. p. 29 in Amiran (1973) – he was a geographer who made this insightful remark on the basis of work in the Middle East, but it is quite apposite to deserts everywhere around the world and nowhere more so than in Australia.

53. Modified after Fig.1 in Stafford Smith (2008).

54. For further discussion, see Stafford Smith (2008).

55. Orians and Milewski (2007).

56. Pickup (1985).

57. Modified after Fig. 5 in Stafford Smith and Morton (1990).

58. These types of patterns are well-described in Ludwig *et al.* (1999); their theoretical rationale is explained by Rietkerk *et al.* (2004).

59. There has been a great deal of work on this idea in the past two decades, see, for example: Pickup (1985); Ludwig *et al.* (1999); HilleRisLambers *et al.* (2001); Rietkerk *et al.* (2004); Ludwig *et al.* (2005); Esteban and Fairen (2006); Dekker *et al.* (2007).

60. Statistics in this section are mostly from Brown *et al.* (2008) and Maru and Chewings (2008). For Canada, see <http://atlas.nrcan.gc.ca/site/english/maps/peopleandsociety/population/population2001/density2001> (accessed March 2009); for Sydney see <http://www.abs.gov.au/AUSSTATS/abs@.nsf/mediareleasesbyReleaseDate/1789776A2CB14348CA256E1C007CE484?OpenDocument> (accessed March 2009).

61. Kindly provided by Vanessa Chewings, CSIRO Alice Springs, from Geoscience Australia topo v3 for settlement locations, and Australian Bureau of Statistics (2001). 'Urban Centre / Locality, Demographic Profile' and CDATA, indigenous statistics, for size of settlements and proportion of indigenous; see also <http://www.desertknowledgecrc.com.au/desert_information/desertmaps.html> (accessed March 2009) for approach.

62. For more discussion, particularly of Aboriginal settlements types in Australia, see Memmott and Moran (2001).

63. For the concept of mobility regions within the desert, see Memmott and Moran (2001) and Taylor (2002).

64. See <http://www.gisca.adelaide.edu.au/projects/aria/> (accessed March 2009).

65. ARIA boundary redrawn from ABS (2001), combining their 'remote' and 'very remote' categories; language data based on AIATSIS/FATSIL (2005); redrawn from Figure 3 in Stafford Smith (2008). The languages are: 1-Alyawarr, 2-Anindilyakwa, 3-Anmatyerre, 4-Arrernte (western Arrernte), 5-Burarra, 6-Dhuwaya (Yolngu), 7-Guugu Yimighirr, 8-Kukatja, 9-Kunwinjku, 10-Maung, 11-Murrinhpatha, 12-Nyangumarta, 13-Pintupi, 14-Pitjantjatjara, 15-Tiwi, 16-Warlpiri, 17-Wik Mungkan, 18-Yulparija.

66. *'Bush Mechanics'* was a four-part television series, shot by director David Batty around Yuendemu in the Tanami Desert, and aired on ABC Television in 2001, which followed the adventures and inventiveness of Aboriginal bush mechanics. They keep cars going in remote places against all odds, but using techniques genuinely applied by people through the region. See <http://www.rebelfilms.com.au/films/bushmechanics/index.html> (accessed March 2009).

67. See keynote address by Mohammed Sherzad (Department of Agricultural Engineering, Ajman University of Science and Technology, United Arab Emirates) on *Building the Sustainability of Desert Communities* at <http://www.desertknowledge.com.au/symposium/program.htm> (accessed March 2009).

68. Language statistics from AIATSIS/FATSIL (2005).

69. See Maru *et al.* (2007); McAllister *et al.* (2008).

70. See the excellent analyses by Tony Smith, e.g. Smith (2006).

71. <http://crh.flinders.edu.au/> (accessed March 2009) and <http://www.crana.org.au/> (accessed March 2009).

72. See Smith (2007). This issue is examined further in Chapters 4 and 9.

73. These strategies, and the additional multi-species ones, are reviewed in Stafford Smith and McAllister (2008), although the 'facilitation' strategy in that source has been more appropriately re-named 'dependents' here. Readers may note that there are animals such as irruptive plains rats and the burrowing frogs that stretch these categories by combining elements of the strategies; and some plant species contain genetic variability that spread risk by some individuals behaving a bit more like one strategy and others like another. Similarly, we shall see, humans come up with other combinations. However, these key strategies capture the major trade-offs that organisms must make in devising whatever combination of strategies they choose.

74. Quote from p. 17 of Ryan and Antoun (2001); but see also Ryan (2004) for further details, including related programs.

75. Flores and Jurado (2003).

76. See Stafford Smith and McAllister (2008) for more details.

77. The two long rainfall records in Europe are for Paris and at Kew Gardens in London. In the 300-year record for Paris, there are only about 10 years (i.e. 3 per cent) below two-thirds of the long-term annual mean of about 600 mm (which has slightly increased since 1688; Slonosky 2002); the equivalent number is about 7 years in about 300 (2 per cent) for Kew Gardens (Wales-Smith 1973; Mayes 2004). In contrast, for Alice Springs 28 per cent of years are below two-thirds of its long-term annual mean of about 250 mm (for the 108 years of quality-controlled data – see > http://www.bom.gov.au/cgi-bin/climate/hqsites/site_data.cgi?variable=rain&area=aus&station=0 15590&period=annual&dtype=raw<, and 3 per cent of years are less than one-third of the long-term annual mean. Paris' wettest and driest years in its 300 year record were 1.5 and 0.45 times its average (901 and 267 mm, respectively), while Alice's wettest and driest of 108 years were 3.1 and 0.23 of its average (782 and 58 mm). A longer record will undoubtedly expand this range.

78. p. vii in Spencer (1896a).

79. p. 9 in Spencer and Gillen (1899).

80. Davidson eloquently discusses how tantalisingly difficult this is on pp. 13–14 in Davidson (2006).

81. Statistics for this paragraph and the next are from Chapter 13 in Keen (2004).

82. E.g. see description of Ngurunderi Dreaming at <http://www.interpretationaustralia.asn.au/infofiles/ Kangaroo%20Island%20-%20stories%20about%20change%20and%20nature.pdf> (accessed March 2009) and Memmott (2005) for northern Australia.

83. For Australia, see <http://mc2.vicnet.net.au/home/cognit/shared_files/cupules.pdf> (accessed March 2009) and <http://www.metmuseum.org/toah/hd/ubir/hd_ubir.htm>; for Europe, see <http://www.metmuseum.org/ toah/hd/lasc/hd_lasc.htm> and <http://www.metmuseum.org/toah/hd/chav/hd_chav.htm> (all accessed March 2009).

84. See Ian Keen's comparisons of Yolngu society with high levels of polygyny and concentrations of power in a few men's hands with more egalitarian (among men, but also between men and women) desert and southern peoples in less resource-rich and reliable areas (Keen 2004).

85. Kunoth-Monks (2006) available at <http://www.desertknowledge.com.au/symposium/program.htm> (accessed March 2009), also transcribed at <http://www.desertknowledge.com.au/symposium/assets/Rose%20Kunoth-Monks.pdf>(accessed March 2009).

86. See the section 'From the Notebooks', pp. 163–206, on nomadic cultures around the world, as well as later sections of quotes (e.g. pp. 237–256, 271–275) in Chatwin (1988).

87. This quote is on p. 81 of Kimber (1986) – Walter Smith's fascinating memoires were recorded by Alice Springs historian, Dick Kimber, during Walter's declining days in Alice Springs' Old Timers old people's home. He describes travelling with Sandhill Bob in Chapter 12, and the intense process of passing the knowledge on to children on p. 78.

88. Walsh (2009).

89. Campbell (2006).

90. Walsh (2009); Latz (2007).

91. Information in this paragraph is drawn from Vaarzon-Morel and Gabrys (2009), p. 7; see also Campbell (2006), p. 38.

92. The foregoing examples are noted in more detail (see pp. 50–51) in Rose (1996).

93. p. 158 in McBryde (2000).

94. p. 2 in Bruce Chatwin's book *The Songlines* (Chatwin 1988).

95. This material is summarised very briefly from the immense and excellent collation of information in Dale Kerwin's PhD thesis (Kerwin 2006), kindly made available to us, which highlights the enormity of the networks and the close links between trade and stories, and the fact that the trade was as much in non-material goods as material (some parts available online at: <http://www4.gu.edu.au:8080/adt-root/uploads/approved/adt-QGU20070327.144524/public/01Front.pdf> [accessed March 2009]).

96. pp. 95-96 in Strehlow (1970). Lutheran pastor Carl Strehlow was based at the Hermannsburg Mission in central Australia from 1894 to his death in 1922 at Horseshoe Bend; he travelled widely and sympathetically through the inland, becoming a noted linguist and anthropologist. His son Ted continued the anthropological and linguistic work and in this piece actually speaks of freshwater lakes, but as gently pointed out by Luise Hercus (Hercus 1985, p. 25) this was a mistranslation and the reliable water sources were actually small, but persistent, soaks.

97. A departure described poignantly on pp. 22–43 in Hercus (1985): the last desert Wangkangurru left the Simpson voluntarily in 1901 and walked south to the Bethesda Lutheran Mission at Killalpaninna.

98. See <http://www.scienceinafrica.co.za/2003/may/san.htm> (accessed March 2009) and Bulletin of the World Health Organization (ISSN 0042-9686) vol. 84 no. 5 Geneva May 2006 (doi: 10.1590/S0042-96862006000500008).

99. Quoted on p. 7 in Maclean (2009).

100. George Cooley was speaking at the 2008 Biennial Lake Eyre Basin Conference in Longreach on 16 September 2008 about the success of his remarkable community and settlement, Umoona, near Coober Pedy in South Australia.

101. Researcher Di James puts it this way in her PhD thesis, which explores how to promote such convergence; James (2005).

102. Wild and Anderson (2007).

103. See <http://www.thewest.com.au/aapstory.aspx?StoryName=393669> (accessed March 2009) citing a Central Land Council press release.

104. AIATSIS/FATSIL (2005).

105. Though, admittedly, damage to many other cultural treasures was not so recognised (e.g. see < http://www.culturenet.hr/default.aspx?ID=23174> [accessed March 2009]).

106. Statistics in this paragraph are from Mitchell *et al.* (2005).

107. Langton (2007).

108. Rothwell (2008).

109. Cited on p. 12 in McDonald (1991) as one of many, sometimes patronising, but often real, examples of genuine recognition.

110. Commissioner of Native Welfare 1966, p. 10, cited in Smith (2006).

111. On this, Tony Smith (2006) cites McLeod D (1984) *How the West Was Lost: The Native Question in the Development of Western Australia*. Port Hedland, self-published.

112. Andreasyan and Hoy (2009): they found that, during 1998–2003, Indigenous mortality was up to nine times higher in remote areas than in the general Australian population, compared with three times higher in outer regional areas (closer to cities) and two times higher in very remote areas. They point out that the fact that rates were lowest in the very remote areas runs contrary to claims that increasing remoteness is simply associated with poorer health status.

113. Rothwell (2007).

114. Anderson (2004). (Note: this pre-dated her election to the NT House of Assembly.)

115. Anderson (2002).

116. See Pawu-Kurlpurlurnu *et al.* (2008), as well as LaFlamme (2007) and <http://www.deet.nt.gov.au/education/stages_of_schooling/middle/support_materials/docs/preparation_teaching_learning/lajamanu_story/understandings_ngurrakurlu.pdf> and <http://www.tracksdance.com.au/html/work_2007_milpirri2.html> (both accessed March 2009).

117. Chapter 10 in Walsh (2009).

118. Turner (2005); also <http://www.iad.edu.au/News/MK%20Turner%20poster%20MR%20doc.pdf> (accessed March 2009).

119. p. 38 James (2005).

120. Kunoth-Monks (2006).

121. p. 347 James (2005).

122. See <http://www.desertknowledgecrc.com.au/watersmart/demonstration/northern-territory.html>; <http://www.desertknowledgecrc.com.au/news/downloads/DKCRC-MR-Napperby-Field-Day.pdf> (both accessed March 2009).

123. See <http://www.anra.gov.au/topics/land/pubs/landuse/landuse-historical.html#sec2_3> (accessed March 2009).

124. p. 82 in Ker Conway (1989).

125. p. 219 in McDonald (1991).

126. p. 50 in Shaw (1995).

127. This box draws on McKeon *et al.* (2004) and Stafford Smith *et al.* (2007); the map is simplified and re-drawn from the latter.

128. See <http://www.kidman.com.au/> (accessed March 2009).

129. Purvis (1986).

130. pp. xxi and 62–63 in Bastin and the ACRIS Management Committee (2008); pp. 117–118 summarise the issues for biodiversity monitoring.

131. The issues in this box are considerably expounded on by Botterill and Fisher (2003) for general drought policy history, Stafford Smith (2003) for effects of different drought instruments in rangelands and the difficulty of defining exceptional circumstances conditions sensibly and fairly, Hennessy *et al.* (2008) for climate change and drought frequency in different regions of Australia, Nelson *et al.* (2008) for possible novel approaches to delivering drought support, as well as the original Drought Review (Drought Policy Review Task Force 1990) and the most recent review by the Productivity Commission (2008).

132. This box is based on publications by Peter Johnston (e.g. Johnston *et al.* 1996) and about AussieGRASS (e.g. Carter *et al.* 2000), as well as personal discussions. See also Stafford Smith *et al.* (2007) for further discussion on why regional knowledge systems are needed.

133. See, for example, Steffen *et al.* (2009).

134. Biograze (2000).

135. This real example is briefly documented on p. 18 of Morton *et al.* (1995).

136. See <http://www.environment.gov.au/indigenous/ipa/index.html> (accessed March 2009). See also <http://www.environment.gov.au/indigenous/ipa/declared/ngaanyatjarra.html> (accessed March 2009) for the Ngaanyatjarra Lands IPA.

137. Re-drawn from <http://www.environment.gov.au/indigenous/ipa/map.html> (accessed March 2009) to show the desert boundaries, within which many of the larger IPAs can be found.

138. Estimates from Peter Whitehead (*pers. comm.*) and Peter Davies (*pers. comm.*).

139. Latz (2007).

140. Statistics from Edwards *et al.* (2008).

141. General material is from Vaarzon-Morel and Gabrys (2009) with quote on p. 5.

142. General material is from Maclean (2009) with quote on p. 8.

143. Edwards *et al.* (2008) and Edwards and Allan (2009).

144. Mr Briscoe has sadly since died, but his family are comfortable for him to be named, on the basis of how far beyond the immediate community his influence has extended; he played a critical and leading role in the partnership described here.

145. More details are described in Evans *et al.* (2009).

146. <http://www.ngapartji.org/> (accessed March 2009).

147. <http://gunya.com.au/> (accessed March 2009) – and data following comes from their report, Gunya Australia (2007), accessible through this site.

148. Redrawn from Holmes (1997) and Campbell *et al.* (2008) and superimposed on the desert boundaries; Campbell *et al.* (2008) illustrate how the classification can be used to think about differentiated service delivery strategies (in this case for health) in each region type.

149. Bush foods are traded through a part Aboriginal company (see for example Robins 2005); the tourism activities are effected through Tilmouth Well Roadhouse –<http://www.tilmouthwell.com/index.htm> (accessed March 2009).

150. See <http://www.obebeef.com.au/> (accessed March 2009).

151. See <http://www.desertknowledgecrc.com.au/watersmart/> (accessed March 2009).

152. Ferguson (2008).

153. Desert Knowledge Australia (2006); for more on approaches to clustering (and the unique geographic scale at which this is occurring in Australia) see Taylor *et al.* (2008), and Ifor Ffowcs-Williams' talk at the Desert Knowledge symposium 2008, recording number 19 available at: <http://www.desertknowledgecrc.com.au/publications/audio.html> (accessed March 2009).

154. Redrawn after map at <http://www.desertknowledge.com.au/dka/index.cfm?attributes.fuseaction=bn_intro> – some smaller centres link in irregularly, and some really only aspire to as yet.

155. See Taylor *et al.* (2008) for more details.

156. Some material in this section is sourced from <http://www.cse.csiro.au/research/nativefoods/>, <www.desertknowledgecrc.com.au> and <http://www.outbackspirit.com.au/aboutus.html> and associated web pages. <http://www.desertknowledgecrc.com.au/education/downloads/Angela-Dennett-Hons-Thesis-Structure-Solanum-Centrale.pdf> (all accessed March 2009) provides a fascinating example of how science can help inform the development of cultivation for the bush tomato.

157. See Taylor *et al.* (2008) for more details.

158. Redrawn from leaflet on <http://www.desart.com.au/aborignalartcentres.htm> (the identity of the art centres may be accessed there); <http://www.aboriginalart.org/index.cfm> (both accessed March 2009) is a similar website dealing with all northern art centres.

159. Cited at <http://www.desart.com.au/aboriginalart.htm> (accessed March 2009).

160. See <http://www.desart.com.au/documents/Morphy_CreatingValue.pdf> (accessed March 2009) and Morphy (2005).

161. The story of Leigh Creek may be found in various local histories, although this is dramatised version! e.g. see <http://www.flinderspower.com.au/history> (accessed March 2009).

162. <http://www.flinderspower.com.au/history>; <http://www.britannica.com/ebc/article-9047680> (accessed March 2009) [Babcock and Brown have now bought this company].

163. McGowan (2002).

164. <http://www.murchison.wa.gov.au/> (accessed March 2009): the total population of the Murchison Shire is only about 160 people.

165. See <http://www.pfes.nt.gov.au/index.cfm?fuseaction=page&p=40&m=22&sm=48&crumb=33> (accessed March 2009).

166. See pp. xii and 92 in Moran *et al.* (2007).

167. Moran *et al.* (2007).

168. See Davies *et al.* (2008) for a good description of the Sustainable Livelihoods approach in an Australian context, and Moran *et al.* (2007) for an analysis in an Aboriginal context.

169. Redrawn after Fig. 1 in Stafford Smith *et al.* (2008).

170. See Rothwell (2008).

171. The discussion here is extended in Stafford Smith *et al.* (2008).

172. See Moran (2008) and Stanley (2008).

173. Grey-Gardner (2008).

174. Seemann *et al.* (2008).

175. Altman (1987).

176. As Warren Mundine has said (ABC TV News, 16/6/07), referring to the then recent NT report on child abuse, 'we can't use the excuse of poverty, as there's lots of poor people in the world and they don't have this problem'.

177. p. 10 in Rose (1996).

178. Draws on <http://www.environment.gov.au/heritage/ahc/publications/commission/books/linking-a-nation/chapter-7.html>, <http://www.heritage.gov.au/cgi-bin/ahpi/record.pl?RNE151>, <http://outbackvoices.com/stuart-of-the-territory/the-barrow-creek-affair> (all accessed March 2009).

179. <http://www.samemory.sa.gov.au/site/page.cfm?u=267> (accessed March 2009).

180. <http://www.marvin.com.au/> (accessed March 2009).

181. <http://www.youtube.com/watch?v=O-MucVWo-Pw> (accessed March 2009).

182. These and other statistics here from the BushLight external evaluation report, at <http://www.bushlight.org.au/media/ann%20reps/bushlight%20evaluation%20report.pdf> (accessed March 2009).

183. <http://www.alicesprings.nt.gov.au/> and <http://www.alicesprings.nt.gov.au/astc_site/community/solar_city> (both accessed March 2009).

184. <http://solacoat.com.au/About-Us-pg7818.html> (accessed March 2009).

185. For water intake figures, see <http://www.mayoclinic.com/health/water/NU00283> (accessed March 2009) (note: people living in hot climates or engaged in physical labour need twice as much as those in cooler climates and desk jobs). See also <http://www.who.int/water_sanitation_health/dwq/nutrientschap3.pdf> (accessed March 2009) and Molden (2007, p. 5) for use per calorie.

186. See <http://www.desertknowledgecrc.com.au/news/downloads/DKCRC-Media-Release-January-5-Beating-Australias-water-losses.pdf> and http://www.desertknowledgecrc.com.au/watersmart/publications/. Many of the results mentioned in this section may be found on <http://www.desertknowledgecrc.com.au/watersmart/> (all accessed March 2009).

187. <http://www.desertknowledgecrc.com.au/news/downloads/DKCRC-Media-Release-January-5-Mustering-cattle-from-the-beach.pdf> (accessed March 2009).

188. This paragraph and the next two draw on <http://www.flyingdoctor.net/> (accessed March 2009).

189. <http://www.assoa.nt.edu.au/> (accessed March 2009).

190. See <http://pre2005.flexiblelearning.net.au/newsandevents/features/featuresaugust05.htm> and report at <http://pre2005.flexiblelearning.net.au/projects/resources/mobile_adult_learning_unit_v2.pdf> (both accessed March 2009).

191. <http://www.npywc.org.au/>; a sense of the diversity of funding sources that have to be knitted together can be obtained from <http://www.npywc.org.au/html/programs.html> (accessed March 2009).

192. E.g. Westbrook *et al.* (2008).

193. Material here particularly based on Pittock (2008). For WorleyParsons, see <http://www.worleyparsons.com/InvestorRelations/ASX/Documents/2008/ASTMarketPresentation.pdf>; for geothermal potential, see <http://

www.pir.sa.gov.au/__data/assets/pdf_file/0004/24979/2007_GIA_Ann_Rept_Australia.pdf>; for DESERTEC, see <http://www.desertec-australia.org/content/twf-1-intro.html>; for Europe's proposals, see <http://www. guardian.co.uk/environment/2008/jul/23/solarpower.windpower> (all accessed March 2009); an analysis of the carbon management value of the European proposal may be found in Monbiot (2006). Barrie Pittock has since drafted a manuscript on this topic.

194. Smith (2007).

195. See Dillon and Westbury (2007), and Desert Knowledge Australia (2008).

196. Chapters 5 (by Larson) and 7 (by Maru and LaFlamme) in Smajgl *et al.* (2009).

197. <www.id.com.au> (accessed March 2009) and Ngaanyatjarra Council (2003).

198. Dollery and Fleming (2005).

199. Adepoyibi (2006).

200. E.g. experience from Nigerian local communities where there were also conflicts between municipal and cultural roles when these were loaded into the one organisation.

201. Approximate boundaries based on <www.nntt.gov.au>, <http://www.tjulyuru.com/facts.asp>, Sanders and Holcombe (2008), <http://www.water.gov.au/maps> [for drainage divisions] and NPY Women's Council <http://www.npywc.org.au/> (all accessed March 2009).

202. Ngaanyatjarra Council (2003); see also various pages in <http://www.tjulyuru.com/default.asp>.

203. Drawn after information on <http://www.tjulyuru.com/Comm.asp> and Ngaanyatjarra Council (2003).

204. See, for example, Lane *et al.* (2004); Farrelly (2005); Paton *et al.* (2005); Lane *et al.* (2005); Larson (2007).

205. Information in this box is mostly based on <http://www.lakeeyrebasin.org.au/archive/pages/page11.html> for the community point of view, and <http://www.aph.gov.au/library/pubs/bd/2000-01/01BD104.htm> (both accessed March 2009) for the legislative viewpoint; see also Larson (2007) for how the natural resource management arrangements differ by state.

206. E.g. see <http://www.lakeeyrebasin.org.au/nrm/nrm.html> (accessed March 2009).

207. <http://www.tsra.gov.au/> (accessed March 2009).

208. Sanders and Holcombe (2008).

209. <http://www.localgovernment.nt.gov.au/> (accessed March 2009).

210. Ostrom (1999); the 'tidiness' quote comes from p. 530.

211. The rules seem esoteric but are worth looking at in the context of the desert drivers. The 'rules about rules' are: (i) 'Boundary rules' – who is involved? What are the geographical boundaries? Do you have to be paying taxes to vote? *May be hard to define in sparsely settled deserts.* (ii) 'Position rules' – capabilities and responsibilities of those in formal positions. Must they be elected or are they just the largest landowners? Must the treasurer have an accounting degree? *Clarity is vital in small communities where there is a risk of nepotism and where individuals commonly fill multiple roles.* (iii) 'Authority rules' – what people in formal positions may, must, or must not do. Must they take action if a member is found to be over-grazing their land? Must written records of decisions be kept and made public? *Clear rights and responsibilities are important when these are often undermined by external 'distant voice' actions.* (iv) 'Scope rules' – what outcomes are allowed, required or forbidden. May the organisation set a rent for land use or only police one set elsewhere? Can the group comment on how the neighbouring group is doing things? *These need to be closely tailored to local desert realities, distant from central authorities.* (v) 'Aggregation rules' – how are individual actions turned into final outcomes? Is there discussion until consensus is reached or must a formal majority vote be taken? *Consensus decision making is often important to maintain trust in small*

communities. (vi) 'Information rules' – what kinds of information are available? Must the sale value of land be in the public domain? Must affected people be consulted? *There is often a lack of information in the desert, and privacy can be an issue in small groups.* (vii) 'Payoff rules' – what costs and benefits result to whom? What penalties apply to non-compliance? What benefits are there for participants who do comply? *Financial and social costs and benefits can be unexpected in sparse populations; and costs of enforcement can be high.* Of course communities often don't formally write down a rule in each category, but they follow social norms (e.g. you don't inform on your neighbour) as if they had; Aboriginal culture codified complex rules, but never wrote them down. The problem is that outsiders can't easily see these implicit or verbally transmitted rules, so when new players come into the system, things tend to go wrong because the rule is not visible. *Yet such turnover of people is common in variable remote regions.* See Ostrom (1999), italics show implications for deserts.

212. In fact she calls these 'nested polycentric governance systems' – a terrible bit of jargon, but one that, due to its significance, we should be hearing more of over the coming decades. See Ostrom (1999) and Marshall (2008).

213. Moran and Elvin (2009).

214. See Measham *et al.* (2007).

215. Hunt and Smith (2007).

216. Desert Knowledge Australia (2008).

217. The organisations are: Eyre Peninsula Natural Resource Management Board, Eyre Regional Development Board and Eyre Peninsula Local Government Association, located close to each other in Port Lincoln (see <http://www.epnrm.sa.gov.au/>, <http://www.eyreregion.com.au/> and <http://www.eplga.sa.gov.au/eplga/index.htm>) (all accessed March 2009).

218. The information in the next two paragraphs was drawn from pages on <http://www.apfc.org/> (accessed March 2009).

219. E.g. Graham Marshall, speaking of the effectiveness of the natural resource management arrangements in Australia; Marshall (2008).

220. Dillon and Westbury (2007).

221. This study concerned Wadeye in the Top End – see Taylor and Stanley (2005).

222. See also Gore (2007).

223. From *Warakurna* on Midnight Oil's 1987 album *Diesel and Dust*.

224. Nicolas Rothwell's (2008) eloquent article about the intellectual breakthroughs and campaigns of Noel Pearson and Marcia Langton expresses this well; alcohol and drug abuse underlie many more painful symptoms of remote Aboriginal condition. Behind these, though, lie an absence of a future, feelings of neglect and the transient welfare attention that create the circumstances in which drugs and alcohol take hold. And, behind these factors lie the desert drivers gone wrong. As Chapter 7 discusses, all layers of the problem need addressing, but addressing the former without considering the latter simply leaves the people facing the same drivers again in a few years time.

225. Kunoth-Monks (2006).

226. ABS (2008).

227. E.g. see <http://www.sciencealert.com.au/opinions/20080309-17885-2.html> (accessed March 2009).

228. E.g. see <http://www.questacon.edu.au/indepth/clever/100_years_of_innovations.html> and <http://www.whitehat.com.au/Australia/Inventions/InventionsA.html> (both accessed March 2009) for other examples, whether agricultural or not, often related to an Australian way of life.

229. See, e.g., Reynolds *et al.* (2007) and Verstraete *et al.* (2008).

INDEX